RAND NATIONAL DEFENSE RESEARCH INSTITUTE

Enhancing Family Stability During a Permanent Change of Station

A Review of Disruptions and Policies

Patricia K. Tong, Leslie Adrienne Payne, Craig A. Bond,
Sarah O. Meadows, Jennifer Lamping Lewis, Esther M. Friedman,
Ervant J. Maksabedian Hernandez

Prepared for Office of the Under Secretary of Defense for Personnel and Readiness (OUSD/P&R)
Approved for public release; distribution unlimited

For more information on this publication, visit **www.rand.org/t/RR2304**

Library of Congress Cataloging-in-Publication Data is available for this publication.
ISBN: 978-1-9774-0052-9

Published by the RAND Corporation, Santa Monica, Calif.

Cover: DanielBendjy/Getty Images.

Preface

Permanent change of station (PCS) moves are a common stressor associated with military life. The 2017 National Defense Authorization Act required the Secretary of Defense to submit a report to the Senate and House Armed Services Committees discussing actions taken by the U.S. Department of Defense (DoD) to enhance stability of military families going through a PCS move. The RAND Corporation was asked to provide assistance with the congressionally mandated report and to conduct a more comprehensive review of the impact of PCS moves on family stability. This report contains RAND's broader analysis, which includes a synthesis of information obtained from existing literature, interviews with subject-matter experts, published tabulations, and secondary data analysis to better understand the types of disruptions faced by military families and the existing programs and policies in place to address these disruptions. This report also contains policy implications and recommendations generated by this analysis.

This report should be of interest to DoD and service branch personnel managers who are looking for ways to enhance family stability during PCS moves. It should also be of interest to policymakers and interest groups who are seeking opportunities to assist and support military families.

This research was sponsored by the Office of the Under Secretary of Defense for Personnel and Readiness and conducted within the Forces and Resources Policy Center of the RAND National Defense Research Institute, a federally funded research and development center sponsored by the Office of the Secretary of Defense, the Joint Staff, the Unified Combatant Commands, the Navy, the Marine Corps, the defense agencies, and the defense Intelligence Community. For more information on the RAND Forces and Resources Policy Center, see www.rand.org/nsrd/ndri/centers/frp or contact the director (contact information is provided on the webpage). Questions or comments about this report should be sent to Craig Bond (cbond@rand.org) or Jennifer Lewis (jlamping@rand.org).

Contents

APPENDIXES

Figures and Tables

Figures

Tables

Summary

One key aspect of military life is frequent relocation, also known as permanent change of station (PCS) moves. According to the U.S. Department of Defense (DoD), one-third of military service members experience a PCS move every year. These frequent moves have the potential to affect *family stability*, a family's ability to operate as a cohesive unit with consistent activities and routines, in both positive and negative ways. Furthermore, there is evidence that these disruptions could impact service member satisfaction with military life and could affect service member and family readiness and resilience for some individuals and families.

The 2017 National Defense Authorization Act required the Secretary of Defense to submit a report to the Senate and House Armed Services Committees discussing actions taken by DoD to enhance stability of military families going through a PCS move. The Office of the Under Secretary of Defense for Personnel and Readiness requested analysis from the RAND Corporation to provide assistance with the congressionally mandated report and to conduct a more comprehensive review of the impact of PCS moves on family stability. This report was generated in response to that request.

Moving Is Disruptive

PCS moves are associated with a broad array of disruptions that impact all members of a military family. Based on feedback from the sponsor, we focused on a subset of disruptions to study further, including spouse employment and service member retention intentions. We also examined the extent to which the untimely receipt of PCS orders is considered problematic and may exacerbate disruptions to family stability. Finally, we examined whether families tend to move together as a unit or in a more staggered fashion (with the service member moving before or after the rest of the family) and whether these circumstances lead to further disruptions.

Using existing literature, published tabulations, and feedback from interviews with subject-matter experts (SMEs), we found evidence that PCS moves can disrupt spousal employment and retention intentions. In particular, the literature contains causal evidence that PCS moves lead to losses in spousal earnings and suggestive evidence that PCS moves result in spousal unemployment, spousal underemployment (e.g., working part time when full time is preferred, being overqualified for the current job), and delays in employment among spouses who need to obtain credentials at the new duty location.

Furthermore, the literature reports suggestive evidence that PCS moves are negatively correlated with service member retention intentions, and these negative correlations may be exacerbated by issues related to the timing of PCS moves in relation to deployments and the untimely receipt of PCS orders. We conducted secondary data analysis using Deployment Life Study data to show that service member military commitment, retention intentions, and satisfaction drop, and spouse financial stress increases, in the period right before a PCS move, which further demonstrates that PCS moves are disruptive.

Published tabulations show that about one-third of service members report at least a moderate problem with a PCS move, and interview analysis demonstrates that issues related to moving during peak season, or summer moves, were a frequently mentioned problem related to timeliness. Our interview data with SMEs identified delays in the release of congressional funding needed to support PCS moves, as well as a lack of connectivity between electronic systems used to generate personnel assignments and logistical aspects of moves, as factors that could contribute to disruptions related to timely receipt of PCS orders.

We did not find evidence on the extent to which families move together as one unit rather than separately with the service member moving before or after the rest of the family. We also did not find any evidence on whether PCS-related disruptions are mitigated or adversely affected by how a family moves.

Our review is based on preexisting data sources and thus reflects the interests and representativeness of the original studies. Future research, especially involving primary data collection about the impact of PCS moves on specialized populations (e.g., more-junior pay grades or dual military couples) and topics of interest, could help fill any potential gaps in current understanding.

Existing Policies, Programs, and Services Cover All Identified Disruptions

Using online searches, data obtained from each service branch, and information from interviews with SMEs, we identified a broad set of programs, policies, and services offered by DoD to address disruptions associated with PCS moves. While the set identified is not necessarily exhaustive, it demonstrates the breadth of what is currently available to military families. We created crosswalks between the list of PCS-related disruptions identified and the set of programs and services available to show that each disruption is identified by at least one existing program. Further, almost all of the disruptions—including household management; spouse employment; service member, spouse, and child psychosocial outcomes; child school involvement and engagement; child care; and military family life—are addressed by multiple different programs. Thus, in our judgment, there is no evidence to suggest a need for new programs, policies, or services.

Policy Opportunities Can Improve the PCS Move Process

While we did not find definitive evidence that new policies, programs, or services are needed to address PCS disruptions, there are opportunities to improve the PCS move process to reduce disruptions, particularly on the front end. Specifically, given survey and interview data related to the timeliness of receiving PCS orders, **we suggest that there is potential for increasing the lead time given to families prior to a PCS move and identify three policy implications to improve the PCS process:**

- **Identify ways to ensure that funding needed to support PCS moves is available when needed.** When a service member is notified of a PCS move determines how much time that service member and the service member's family have to plan their move. Notification is directly tied to funding. The earlier funding is released by Congress, for example, the sooner service members can be notified of a move and start planning their relocation. Thus, DoD should consider any and all programming and budgeting actions that would functionally increase lead time for service members facing a PCS move.
- **Improve the demand signal for logistical aspects of PCS moves to mitigate disruptions.** Our interview analysis suggests that improving the demand signal (i.e., the indicator that a move is imminent for those involved in the logistics chain) is particularly important for moves occurring during peak season. A better identification of when the move is going to occur would allow families to get a head start on securing movers, establishing transportation, finding housing, and managing other logistical aspects of relocation.
- **Sync personnel assignment, pay, and PCS systems electronically.** Currently, the service branch systems that process service member assignments and those systems that initiate the logistics required for a PCS move (e.g., the Defense Personal Property System) are not able to communicate with each other. This lack of coordination may cause delays in planning for moves. We suggest instituting an automatic notification system that would send an electronic alert to the U.S. Transportation Command to start the move process. This notification system would likely need to work through a financial system (i.e., personnel actions would impact pay actions and trigger financial transfers, signaling the start of the

logistical PCS process). The Marine Corps Total Force System and the Army's Integrated Personnel and Pay System are current and future mechanisms, respectively, where such notification could be tested.

Understanding Program Efficiency and Effectiveness Requires Additional Research

The effectiveness of existing programs and services in alleviating PCS-related disruptions is not well understood. Further work is needed in this area. Additional research should evaluate the relative effectiveness of different types of service provision mechanisms (e.g., online, call centers, in person) and determine why certain families participate in existing programs while others do not. We suggest that DoD and the service branches collect data on the extent to which service members and families move together or at different times to determine whether either approach mitigates or exacerbates PCS disruptions and to evaluate proposals designed to give families more flexibility along this dimension.

Acknowledgments

We thank Stephen B. Nye, assistant director of Military Personnel Assignments/PCS Policy in the Office of the Under Secretary of Defense for Personnel and Readiness. We also thank Vickie LaFollette, Office of Military and Family Readiness Policy, and Patricia Mulcahy, director of Officer and Enlisted Personnel Management in the Office of the Deputy Assistant Secretary for Military Personnel Policy. We are grateful to the military personnel subject-matter experts who participated in the interviews conducted by RAND. We wish to thank Robin Beckman for assistance with constructing tabulations with the Deployment Life Study data and Barbara Bicksler for her expert help in improving the clarity of the exposition. Finally, we also thank Tom Trail and Shelley MacDermid Wadsworth for carefully reviewing this report.

Abbreviations

A&FRC	Airman and Family Readiness Center
ACS	Army Community Center
ADSS	Survey of Active Duty Spouses
BAH	basic allowance for housing
COE	Continuity of Education
CONUS	continental United States
COT	Consecutive Overseas Tour
DLS	Deployment Life Study
DMDC	Defense Manpower Data Center
DoD	U.S. Department of Defense
DPS	Defense Personal Property System
DTIC	Defense Technical Information Center
DTMO	Defense Travel Management Office
EFMP	Exceptional Family Member Program
FFSC	Fleet and Family Support Center
FY	fiscal year
GAO	U.S. General Accounting Office
HEAT	Housing Early Assistance Tool
IPCOT	In-Place Consecutive Overseas Tour
IPPS-A	Integrated Personnel and Pay System–Army

L.I.N.K.S.	Lifestyle Insights, Networking, Knowledge, and Skills
MCCS	Marine Corps Community Services
MCTFS	Marine Corps Total Force System
MSEP	Military Spouse Employment Partnership
MyCAA	My Career Advancement Account
OCONUS	outside the continental United States
OMB	Office of Management and Budget
OTEIP	Overseas Tour Extension Incentive Program
OUSD(AT&L)	Office of the Under Secretary of Defense for Acquisition, Technology, and Logistics
OUSD(P&R)	Office of the Under Secretary of Defense for Personnel and Readiness
PCS	permanent change of station
SECO	Spouse Education and Career Opportunities
SME	subject-matter expert
TO	Transportation Office
TOS	Time-on-Station
TRANSCOM	U.S. Transportation Command
USCG	U.S. Coast Guard

CHAPTER ONE

Introduction

One key aspect of military life is frequent relocation, also known as permanent change of station (PCS) moves. According to the U.S. Department of Defense (DoD), approximately one-third of military service members experience a PCS move every year. In 2016, 71 percent of active-duty members report having made at least one PCS move over the course of their service (Defense Manpower Data Center [DMDC], 2017). On average, the number of PCS moves reported by active-duty service members is 2.6, with 31 percent reporting one move and 15 percent reporting six or more moves (DMDC, 2017).[1] These moves tend to disrupt *family stability*, a family's ability to operate as a cohesive unit, in negative ways; however, positive effects (e.g., moving to a more desirable location) are also possible. These disruptions could potentially impact service member satisfaction with military life, thereby influencing future retention decisions and affecting military readiness and resilience. For example, the 2017 Blue Star Families Military Family Lifestyle Survey reports that 44 percent of service members and 45 percent of spouses indicate "relocation stress" as one of the top five stressors experienced in their career as a military family (Shiffer et al., 2017). Although the Blue Star report is not a statistically representative sample, it indicates that PCS moves are clearly viewed as problematic by many military families.

The 2017 National Defense Authorization Act required the Secretary of Defense to submit a report to the Senate and House Armed Services Committees discussing actions taken by DoD to enhance the stability of military families going through a PCS move. The Office of the Under Secretary of Defense for Personnel and Readiness (OUSD[P&R]) requested analysis from the RAND Corporation to provide assistance with the congressionally mandated report and to conduct a more comprehensive review of the impact of PCS moves on family stability.

Our review is based primarily on existing literature, published tabulations, and feedback from interviews with subject-matter experts (SMEs).[2] More specifically:

1 Average is reported over all service member respondents at the time of the survey, conditional on their time of service.

2 The study was reviewed by RAND's Human Subjects Protection Committee and was determined not to be human subjects research.

- *To identify disruptions to family stability generated by PCS moves*, we used a mixed-methods approach, which included leveraging team member knowledge, consulting with RAND Corporation military fellows (e.g., midlevel officers in the Air Force and Army), and reviewing existing literature and data sources.
- *To conduct in-depth reviews of select disruptions to family stability*, we performed a focused literature review, interviewed high-ranking officials with subject-matter expertise in the area of PCS moves, used published statistics from DMDC surveys, and performed secondary data analysis, when possible. The three disruptions selected by the study sponsor for in-depth review were (1) disruptions to spouse employment, (2) disruptions to retention intentions, and (3) disruptions associated with moving as a family unit versus moving at different times.
- *To identify existing DoD and service programs addressing PCS moves and the associated disruptions*, we conducted an online search of resources available to service members and families, including Military OneSource and service branch–specific websites; reviewed information from the service branches provided by the sponsor; and held interviews with SMEs across all DoD service branches and the Coast Guard.

For the literature review component of the research, we used traditional academic databases, such as JSTOR and Google Scholar, to search for peer-reviewed publications. We also searched the Defense Technical Information Center (DTIC), which houses various collections of military-related documents, including scientific and technical reports from defense-sponsored research. In addition, we searched through publications from RAND and the U.S. General Accounting Office (GAO) (now known as the U.S. Government Accountability Office). Keywords used in these searches included "permanent change of station moves," "PCS," "frequent relocation," "military spouse employment," "children school mobility," "parental absences," and "military divorce."

This report contains the results of our review and is organized as follows. In Chapter Two, we provide a model of disruptions to the stability of military families associated with PCS moves and a more detailed examination of specific disruptions and issues related to PCS moves based on feedback from DoD. In particular, we compile findings about disruptions related to spouse employment, retention intentions, and moving together as a family unit rather than separately (with the service member moving before or after the rest of the family). In Chapter Three, we identify existing policies, programs, and services that address the disruptions identified in Chapter Two, including those offered by DoD that are available to all active-duty military personnel and to service members in select service branches. In Chapter Four, we discuss policy implications and offer recommendations for future research to better understand the role of current programs in mitigating PCS move disruptions.

This report also contains four appendixes, which provide greater detail about interviews with SMEs (Appendix A), the analysis of Deployment Life Study data

(Appendix B), existing programs and services designed to address service member and family needs during a PCS (Appendix C), and the service branches' assignment policies related to PCS moves (Appendix D).

CHAPTER TWO

Permanent Change of Station Move Disruptions

This chapter contains an overview of disruptions to the stability of military families associated with PCS moves. We begin by taking a broad look at the range of possible disruptions and then engage in a more focused review of published evidence about select disruptions, including spouse employment and retention intentions. Subsequent sections address two aspects of PCS-related disruptions: how retention-related outcomes may change over a PCS cycle and the timing between PCS moves and deployment. Finally, we examine whether military families tend to move together as a single unit or separately (with the service member moving before or after the rest of the family) and whether the moving mode is associated with further disruptions.[1] The chapter concludes with a summary of the results.

Types of Disruptions

PCS moves generate disruptions to family stability in direct and indirect ways, as illustrated by the evolution of the PCS move cycle modeled in Figure 2.1. The PCS move cycle begins, as shown at the top of the figure, when a service member receives PCS move orders. Next, the family must decide when and where the spouse and children will move, and, finally, the service member and the family (if they decide to move) complete their relocation. Associated with these moves are both first-order and second-order disruptions. First-order disruptions are ones that are a direct consequence of the PCS move, such as moving household goods, finding housing at the new duty location, and changing schools. These first-order disruptions generally occur at any time between when the PCS move orders are received and when the service member and family complete their relocation. Second-order disruptions are an indirect consequence of the PCS move—they are a byproduct of the move and do not necessarily occur during the PCS move cycle itself. The need for family members to build new

[1] While our review of disruptions is intended to illustrate the potential disruptions across all families, it is not the case that every family (or individuals within that family) will experience the same disruptions in the same way.

Figure 2.1
Permanent Change of Station Move Cycle

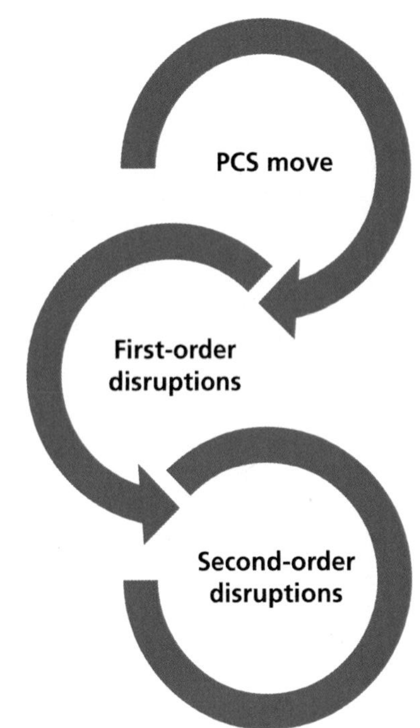

- Service member is assigned to new duty station via PCS orders.
- Spouse and/or child(ren) decide when and where to move relative to the service member.
- Move is completed by the service member and/or the family.
- DoD and the armed services have policies, programs, and services to address disruptions associated with PCS moves.

- First-order disruptions are problems or issues that the service member, spouse, and child(ren) experience as a **direct** consequence of a PCS move (e.g., finding housing at a new location, enrolling in a new school).
- First-order disruptions may occur any time between the receipt of PCS orders through the move itself.
- Policies, programs, and services that address first-order disruptions are tied to the PCS cycle.

- Second-order disruptions are problems or issues that the service member, spouse, and child(ren) experience indirectly as a byproduct of a PCS move (e.g., family functioning).
- Policies, programs, and services that address second-order disruptions may not be tied to a PCS cycle, as these issues may occur as a result of problems not tied to a PCS move.

NOTE: Although this cycle is conceptualized as a single move, problems may compound over multiple moves.

RAND *RR2304-2.1*

social networks and the effect of frequent moves on retention intentions are examples of second-order disruptions.

Our mixed-methods approach revealed a number of disruptions that affect the service member, spouse, children, and family as a whole (see Table 2.1). The listed categories by family member type are not necessarily mutually exclusive. For service members themselves, household management is a first-order disruption. Household management includes financial costs (i.e., things not reimbursed by DoD), dealing with the logistics of the move itself (e.g., availability of offices designated to help with PCS), selling or renting a prior residence, setting up a new household, and arranging for temporary or installation housing (if desired). Service members may also experience second-order disruptions to military life and psychosocial outcomes. Disruptions to military life include problems related to job satisfaction and retention intentions. Psychosocial outcomes cover mental health problems, substance use and abuse, and difficulties related to social integration and peer support (e.g., developing new friends).

For military spouses, household management and employment are considered first-order disruptions. Employment-related disruptions include a reduction in hours

Table 2.1
Disruptions Associated with PCS Moves

Family Category	First-Order Disruptions	Second-Order Disruptions
Service member	Household management • Cost • Logistics of the move itself • Selling or renting old residence • Setting up a new household • Availability of temporary or installation housing	Military life • Job satisfaction • Retention intentions Psychosocial outcomes • Mental health issues • Substance use • Social integration • Peer support • Other behavior problems
Spouse	Employment • Hours worked and income • Credentialing and licensing • Educational attainment Household management • Cost • Logistics of the move itself • Selling or renting old residence • Setting up a new household • Availability of temporary or installation housing	Military life • Satisfaction with military life • Retention intentions Psychosocial outcomes • Mental health issues • Substance use • Social integration • Peer support • Other behavior problems
Child	Changing schools • Enrolling • Grade retention	Psychosocial outcomes • Mental health issues • Substance use • Social integration • Peer support • Other behavior problems School performance and engagement • Attendance • Test scores • Grades • Homework completion
Family	Child care • Cost • Location • Availability • Quality	Family functioning • Marital satisfaction and quality • Marital stability • Communication • Quality of parent-child relationship

NOTE: The listed categories by family member type are not necessarily mutually exclusive.

worked, decreased income or underemployment, and problems related to location-based credentialing and licensing and educational attainment (e.g., transferring credits). PCS moves can also create the same set of household management disruptions as those listed for the service member. Further, spouses may experience the same set of psychosocial effects as those identified for the service member, as well as changes in spousal satisfaction with the military and spousal preferences over reenlistment intentions.[2]

First-order disruptions for military children include those associated with changing schools. Relocating will cause families to have to enroll their children in new schools and, depending on whether the new school's requirements are more or less strict than those at the old school, could directly impact which grades their children are in. We consider other measures of school performance and engagement (e.g., attendance, test scores, grades, homework completion) to be second-order outcomes that could occur outside of the PCS move cycle itself. PCS moves could also lead to difficulties with children's psychosocial outcomes, including mental health issues, behavior problems, substance use and abuse, and social integration and peer support. For example, PCS moves have been linked to spouse stress, child stress, and child disruptions associated with changing schools and peer groups (Clever and Segal, 2013; Drummet, Coleman, and Cable, 2003; Morgan, 1991; U.S. Chamber of Commerce, 2017).[3]

At the family level, PCS moves could impact access to child care, which we categorize as a first-order disruption because relocation will require parents to find child care at the new duty location. Second-order disruptions at the family level include problems related to family functioning, such as lowered marital satisfaction and quality, marital instability, poorer marital communication, and disruptions to the parent-child relationship.[4]

Although Table 2.1 contains a lengthy list of disruptions to family stability created by PCS moves, we point out three important caveats. First, there are potentially positive effects from frequent location, such as promoting a service member's career progression and increasing family member resilience (e.g., Spencer, Page, and Clark, 2016). In our interviews with SMEs, the most common positive aspect of PCS moves mentioned was increased readiness and resilience, particularly among children. In

---

[2] Although an in-depth examination of disruptions faced by dual-military (or mil-to-mil) couples is beyond the scope of this study, it is important to note they may face a unique set of challenges. Interview respondents described a commensurate yet structurally different variety of stress experienced by dual military couples who experience PCS moves. As one interviewee described, "There's definitely a different dynamic—I don't know if it's harder or easier, though. For mil-to-mil families, we're now trying to manage their careers so they are upwardly mobile, while at the same time detailing them as much as we can to the same geographic location—that doesn't necessarily mean that they are in the same city. The commute between two locations may be horrendous, but they're in the same geographic area."

[3] A full exploration of child-related outcomes is beyond the scope of this study. For further details, please see the cited literature.

[4] These second-order family disruptions can be exacerbated by the peculiarities of any particular moving experience or, more generally, by market and other conditions. For example, housing shortages might result in a given family moving multiple times in the same regional location, leading to disruptions in school performance and social networks. Similarly, some moves might not require any change in installation or school, reducing the disruptions associated with moving. We thank a reviewer for pointing out these examples.

addition, some military families might have a preference for frequently relocating and enjoy the challenges associated with moving to new areas.

Second, the existence and potential deleterious effects associated with any given disruption identified in Table 2.1 depend on many different factors, including one's previous experience with PCS moves; the structure of a service member's family, including the service member's marital status and number and age of children; and the presence of a family member requiring special medical or educational needs. Furthermore, the disruptions associated with a change in household location are not necessarily unique to military families, although the PCS process virtually ensures exposure to a move and the potential disruptions.

Third, while most of the evidence base deals with the effects of a single move, it is possible that multiple moves can exacerbate the first- and second-order disruptions. More research is needed to fully understand the effects of multiple moves.

Based on feedback from the study sponsor, we conducted in-depth research into PCS-related disruptions in three areas: (1) spouse employment, (2) retention intentions, and (3) moving together as a family unit versus moving separately at different times. We discuss each of these issues in the following sections.

Spouse Employment

Problems related to spouse employment were the second–most frequently mentioned negative aspect of PCS moves, according to our interviews.[5] To further understand the types of problems spouses experience with respect to employment after a PCS move, we turned to relevant literature and published tabulations from existing surveys.

PCS Moves Can Contribute to Disparities Between Desired and Achieved Employment Outcomes

There is evidence to suggest that military spouses do not achieve their desired job outcomes. For example, the DMDC 2015 Survey of Active Duty Spouses (ADSS) is a representative survey of roughly 9,800 active-duty spouses that solicits information on a broad set of topics, including employment and PCS moves.[6] Published statistics from the ADSS demonstrate that there is a gap between desired and realized employment outcomes among military spouses. Forty-one percent of active-duty spouses reported

[5] Needs related to having a family member with an Exceptional Family Member Program (EFMP) were the most frequently mentioned negative aspect of PCS moves during our interviews. Although this issue is clearly important for those service members and families affected, it represents a specialized need outside our scope and should be addressed through additional, focused study.

[6] Spouses of National Guard and Reserve service members are excluded from the DMDC 2015 ADSS survey.

being employed, yet 58 percent of military spouses reported needing to work, and 82 percent reported wanting to work.[7]

In addition, there is evidence that some military spouses are underemployed, meaning that they are overqualified for their jobs or want to work full time but work part time. Rates of underemployment among military spouses are generally higher than among nonmilitary civilian spouses. For example, according to a RAND study by Lim and Schulker (2010) that used March 2006 Current Population Survey data, 38 percent of military wives had relatively high levels of education for their current jobs. That rate was markedly higher than the corresponding rate for civilian wives (6 percent). The disparity in educational attainment and job mismatch could reflect employment obstacles unique to military spouses, such as frequent relocation, or could reflect differences in preferences for types of jobs. Published results from the 2015 ADSS showed that just over half of military spouses reported being currently employed in their area of education or training.[8] This could be evidence of underemployment, or it could reflect spouses choosing to work at jobs that are not in their field of expertise for other reasons.

Focusing on published results that might better reflect being involuntarily underemployed, Lim and Schulker (2010) found that 9 percent of military wives worked part time but would prefer to work full time. This rate is much higher than the rate reported by civilian wives (2 percent). Further demonstrating that some military spouses are involuntarily underemployed, the 2015 ADSS showed that 17 percent of military spouses reported that the main reason for working part time instead of full time was that they could find only part-time work.[9]

Three more-recent studies used surveys to assess issues related to military spouse employment. The first is a 2014 study by Maury and Stone, which is based on a 2013 survey of active-duty spouses. The second is a 2017 study by the U.S. Chamber of Commerce's program *Hiring Our Heroes*, which surveyed active-duty military service members and veterans who had retired or left the military within the past three years. The third is the 2017 Blue Star Families Military Family Lifestyle Survey (Shiffer et al., 2017). All three studies rely on convenience samples, meaning that their surveys are not necessarily representative of the military spouse population. Nevertheless, they offer useful insights. Maury and Stone (2014) found that 28 percent of military wives were not working because they could not find work that matched their skills or education, suggesting that a sizable portion of nonworking spouses is involuntarily unemployed. The U.S. Chamber of Commerce study reported that 18 percent of employed

---

[7] These statistics are restricted to spouses responding to these survey items.

[8] This statistic is restricted to spouses who answered the relevant survey item and were employed or currently serving in the military.

[9] This statistic is restricted to spouses who answered the relevant survey item, were employed, and reported working fewer than 35 hours per week.

military spouses had seasonable or temporary jobs, of which 82 percent would prefer a permanent position instead, providing further evidence of underemployment. Finally, the 2017 Military Family Lifestyle Survey found that 55 percent of military spouse respondents were underemployed (i.e., were overqualified, underpaid, or underutilized for their current position), and 41 percent were currently earning less than half of their previous highest salary. Even among military spouses who are employed, earnings may be stifled by the military lifestyle, including the need to relocate every few years. While suggestive, these findings are not direct evidence that PCS moves are causally related to employment outcomes.

PCS Moves Lead to Losses in Spousal Earnings

A body of existing research demonstrates the existence of differences in employment outcomes between military and civilian spouses. In particular, research shows that the probability of being employed and the probability of having average earnings are both lower among military spouses relative to civilian spouses (Burke and Miller, 2017; Heaton and Krull, 2012; Hosek et al., 2002; Lim and Schulker, 2010; Lim, Golinelli, and Cho, 2007; Meadows et al., 2016). These studies generally use a combination of administrative military personnel data and survey data in which the analyses are conducted by comparing military spouses with civilian spouses while controlling for differences in demographic characteristics. Frequent moves are often cited as a reason for these differences in employment outcomes between military and civilian spouses.

A 2017 article by Burke and Miller is the only study, to our knowledge, that estimates a causal relationship between PCS moves and spousal employment. That study used longitudinal administrative military personnel data merged with Social Security Administration Form W-2 wage earnings data for years 2001–2012. The authors found that a PCS move caused an average loss in spousal earnings of 14 percent, or about $3,100, in the calendar year of the move for working spouses. They also found that, on average, a PCS move increased the likelihood that a spouse had no earnings in the calendar year of the move. Older spouses, those with young children, and male spouses experienced larger wage earnings losses, on average, as a result of a PCS move. Importantly, the loss in spousal earnings was persistent, with significant differences remaining two years after the move.

A Majority of Spouses in the Labor Force Report Problems with Finding a Job Following a PCS Move

Published tabulations from the 2015 ADSS provide further evidence that PCS moves disrupt spouse employment. For example, there are large fractions of employed and unemployed spouses who reported problems finding a job following their most recent PCS move. Figure 2.2 shows that 60 percent of employed military spouses and 81 percent of unemployed military spouses reported that finding a job was at least a moderate

Figure 2.2
For the Most Recent PCS Move, to What Extent Was Finding a Job a Problem?

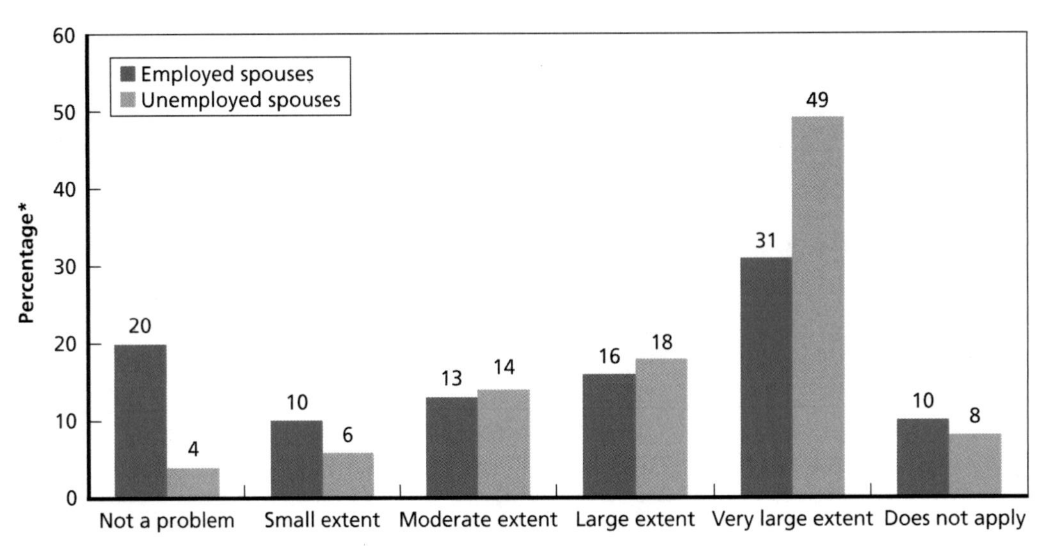

SOURCE: DMDC, 2015.
NOTE: Some percentages do not total 100 because of rounding.
* Percentage of respondents who answered the relevant survey item and had experienced at least one PCS move at any point in the past.
RAND *RR2304-2.2*

problem.[10] The percentages of employed and unemployed spouses who reported that finding a job was a very large problem were 31 percent and 49 percent, respectively. The 2015 ADSS also presents statistics on the number of months it took for military spouses to find employment after their most recent PCS move. Thirty-eight percent of spouses took at least seven months to find a new job after their last PCS move, and 38 percent of spouses took less than four months to find a new job, as shown in Figure 2.3.[11]

Spouses Requiring Credentials to Work Can Experience Delays in Employment

Military spouses are commonly employed in occupations that require location-based credentials to work. In 2012, the Treasury Department and DoD conducted a study demonstrating the need to streamline occupational licensing for military spouses to facilitate employment opportunities when they move across state lines. Using survey data from the 2007–2011 Annual Social and Economic Supplement of the U.S. Census Bureau's Current Population Survey, which included information from about

---

[10] These statistics are restricted to those who answered the relevant survey item and had experienced at least one PCS move at any point in the past.

[11] These statistics are restricted to those who answered the relevant survey item, had experienced at least one PCS move at any point in the past, and indicated that the question was applicable.

Figure 2.3
How Long Did It Take to Find Employment After the Last PCS Move?

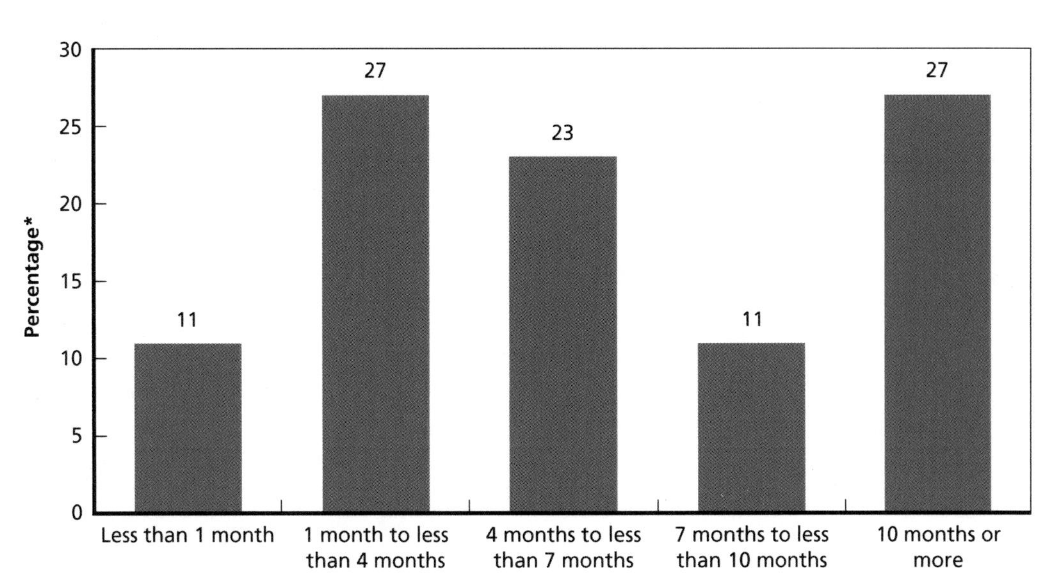

SOURCE: DMDC, 2015.
NOTE: Percentages do not total 100 because of rounding.
* Percentage of respondents who answered the relevant survey item, had experienced at least one PCS move at any point in the past, and indicated that the item was applicable.
RAND *RR2304-2.3*

2,800 military spouses, the study found that the top three self-reported occupations for those in the labor force were teacher, child care worker, and registered nurse—all occupations that require a license or certification. (Note that these figures apply to military spouses in the labor force, which means that both employed and unemployed spouses are included.)

The 2015 ADSS asked working spouses about the career field of their current job. The top three most prevalent categories were "other unlisted occupations not requiring a state license," health care/health services, and education.[12] Certain jobs within health care/health services and education require location-based credentials, such as nursing and teaching. According to the 2015 ADSS, 35 percent of all military spouses, regardless of their employment status, reported being in an occupation or career field that requires a state-issued license.[13] In addition, 27 percent of military spouses reported

[12] These statistics were restricted to those who answered the relevant survey item and reported being employed. Other career field options included information technology, financial services, recreation and hospitality, administrative services, child care/child development, animal services, skilled trades, and communications and marketing.

[13] This statistic is restricted to those who answered the relevant survey item.

having to acquire a new professional/occupational license or credential to work at the new duty location after their most recent PCS move.[14]

There is evidence that the career mismatch among military spouses is higher among service industry–related jobs. Maury and Stone (2014) asked respondents to identify a career field for their current or most recent job and then asked respondents whether their current or most recent job was in their preferred career field. Retail/customer service, hospitality, child care, and administrative services were the career fields with the **lowest** percentage of military wives indicating that this was their preferred field. Health care, education, and government were the career fields with the **highest** percentage of military wives indicating that this was their preferred field. However, caution should be used when interpreting these results: Attributing any career mismatch to PCS moves specifically may inflate the association because some military wives might choose to work outside of their preferred career field for other reasons. However, the results are suggestive of the fact that frequent moves mean that military spouses might have to accept available low-skilled work that generally does not require credentialing or licensing but that may not be the best fit in terms of preference.

Other evidence suggests that PCS moves can lead to problems associated with job-related credentials, education, and training for spouses. According to the 2015 ADSS, among military spouses who indicated that they would like to be in school or training, 28 percent reported that moving too often prevented them from attending school or training.[15] Forgoing schooling and training could prevent spouses from attaining licenses and certifications and could also prevent spouses from pursuing their most preferred careers. Using published tabulations of the 2015 ADSS, Figure 2.4 depicts the number of months it took for military spouses to acquire a new professional or occupational license or credential in order to work at the new duty location after their most recent PCS move.[16] While half of spouses took less than four months to acquire a new credential, 30 percent took at least seven months, and of those, 22 percent took ten months or more.

Some Spouses View Being a Military Spouse as Potentially Detrimental in the Labor Market

Some studies report that spouses believe that military affiliation makes it more difficult to find employment (Castaneda and Harrell, 2008; Harrell et al., 2004; Maury

[14] This statistic is restricted to those who answered the relevant survey item, had experienced at least one PCS move at any point in the past, and indicated that the question was applicable.

[15] This statistic is restricted to those who answered the relevant survey item and indicated that they would like to be in school or training.

[16] These statistics are restricted to those who answered the relevant survey item, had experienced at least one PCS move at any point in the past, and reported having to acquire a new professional or occupational license or credential to work at their new duty location.

Figure 2.4
How Long Did It Take to Acquire a New Professional or Occupational License or Credential in Order to Work at the New Duty Location?

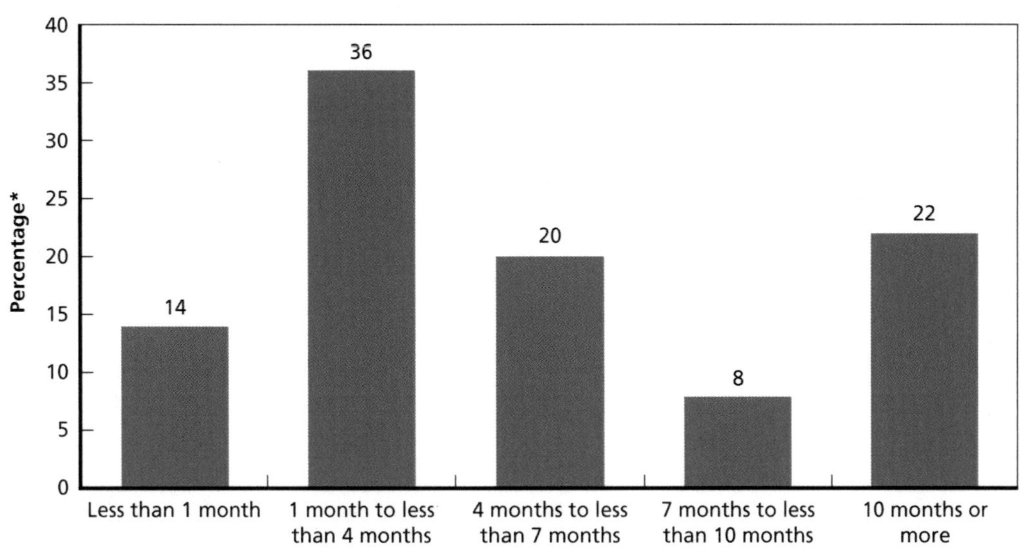

SOURCE: DMDC, 2015.
* Percentage of respondents who answered the relevant survey item, had experienced at least one PCS move at any point in the past, and reported having to acquire a new professional or occupational license or credential to work at their new duty location.
RAND *RR2304-2.4*

and Stone, 2014; U.S. Chamber of Commerce, 2017). For example, a RAND study by Harrell et al. (2004), in which interviews with military spouses were conducted, reported that almost 10 percent of spouses perceived negative stigma among employers and cited it as a barrier to employment. Interviewees commented that employers did not want to hire people who would leave eventually and that, even if hired, employers may not offer the same opportunities for training and advancement because they view military spouses as temporary employees.

More-recent studies by Maury and Stone (2014), the U.S. Chamber of Commerce (2017), and Blue Star Families (Shiffer et al., 2017) demonstrate that spousal beliefs about military stigma still exist. Maury and Stone (2014) found that 59 percent of military wives said that they would not tell a potential employer they were military spouses, with the top-cited reason being that they thought it would make an employer less likely to hire them. The U.S. Chamber of Commerce (2017) study asked military spouses to select the two or three greatest challenges for military spouses seeking good employment opportunities from a list of potential challenges. Forty-one percent of military spouses surveyed listed "The company doesn't want to hire a military spouse because they may move" as one of the top challenges. The Blue Star Families 2017 Military Family Lifestyle survey found that 77 percent of spouse respondents believed that being a military spouse had negatively impacted their careers. To the authors' knowledge, there is no objective evidence in the literature that supports or refutes these perceptions from the standpoint of employers.

Retention Intentions

PCS moves impact service member retention intentions both directly, based on service member satisfaction with the military, and indirectly, as a result of spouse stressors and spouse satisfaction with military life. Our analysis also shows that the timing of PCS moves and deployments and delays in receiving PCS orders could exacerbate these relationships.

Evidence That PCS Moves Are Correlated with Service Member Retention Intentions

Information from the existing literature and from our own secondary data analysis using data from the Deployment Life Study (DLS), a longitudinal study of married military families over the deployment cycle (pre-deployment, during deployment, and post-deployment), provide evidence that PCS moves are correlated with service member retention intentions.[17] The evidence presented in this section is not causal, meaning that we cannot quantify changes in retention intentions attributed to PCS moves. Furthermore, the outcome of interest in this section is retention intentions, not realized retention, and retention intentions may not be perfectly correlated with actual retention rates. To facilitate the discussion, we summarize our findings in Figure 2.5. It depicts the different paths connecting an increase in the frequency of PCS moves and service member retention intentions. The figure is based on the existing literature and RAND DLS analysis detailed below.

Evidence That PCS Moves Directly Impact Retention Intentions

The literature suggests that the frequency of moves is negatively associated with service member satisfaction, commitment to military life, and retention intentions. For example, the U.S. Chamber of Commerce (2017) study reported that 28 percent of military spouses cited frequent relocations as a very important factor that affected whether the service member stayed in the military.[18] In addition, estimates from a previously published RAND DLS Study (Hengstebeck et al., 2016) demonstrated that frequent relocation was negatively and statistically significantly correlated with service members' military satisfaction.

[17] For more information on the DLS, see Tanielian et al. (2014) and Meadows, Tanielian, and Karney (2016).

[18] "Very important" was defined as a ranking between 8 and 10 on a 1- to 10-point scale, with 10 being the most important.

Figure 2.5
Pathways of Correlation of PCS Moves with Service Member Retention Intentions

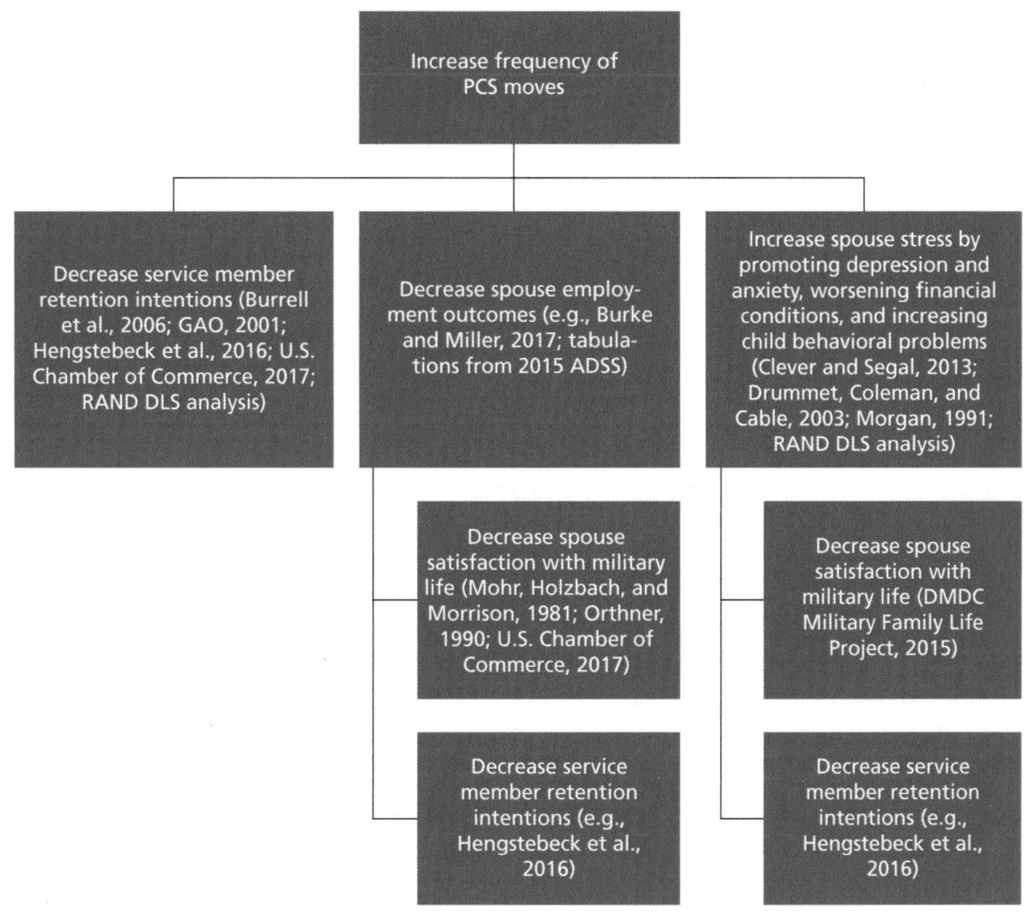

NOTE: Citations refer to evidence linking the outcomes in the current box to the outcomes in the previous box.
RAND *RR2304-2.5*

A 2001 GAO report further demonstrated that the frequency of moves was negatively correlated with service member retention intentions and satisfaction using data from DoD's 1999 Survey of Active Duty Personnel. For example, GAO found that, across most service branches, about half or more of service members averaged less than two years between moves. Service members averaging less than two years between moves reported that they were less likely to stay in the military and reported lower satisfaction with military life than did service members averaging at least two years between moves. Focusing specifically on the Army, Burrell et al. (2006) estimated that the impact of moving was negatively and significantly correlated with Army life satisfaction.[19] That study was based on a survey conducted in 2002 in which questionnaires were sent out to units stationed in Germany. While these results may not be generalizable to the broader active-duty service member population, the paper provides some evidence that moving is directly and negatively correlated with satisfaction with military life.

Evidence That PCS Moves Indirectly Impact Retention Intentions

The literature suggests that spousal satisfaction with military life is associated with stressors, including, but not limited to, the frequency of PCS moves and lack of employment, and that, ultimately, spousal satisfaction is likely to be associated with service member retention intentions. Therefore, the need to relocate every few years may not be statistically tied to retention outcomes but, rather, may serve as one of many factors that affect attitudes about and satisfaction with the military. These factors could then subsequently translate into decisions about staying in the military (see Figure 2.5).

The RAND DLS study (Hengstebeck et al., 2016) estimated the relationship between frequent relocation and spouse preferences. It found negative, but only marginally statistically significant, associations between frequent relocation and spouse military satisfaction and military commitment. The 2001 GAO study also provides some additional evidence that the frequency of moves is negatively correlated with spousal preferences about service member retention. In particular, for service members averaging less than two years between moves, a higher percentage of spouses and significant others were in favor of the service member leaving the military.

As documented earlier, there is evidence that PCS moves negatively impact spousal employment outcomes. The existing literature also provides evidence that working spouses tend to have less favorable views about retention (Mohr, Holzbach, and Morrison, 1981) and that unemployed spouses who were looking for work were the least satisfied with military life (Orthner, 1990). The U.S. Chamber of Commerce (2017) study asked spouses, "Thinking about some factors that people have told us affected whether or not their spouse stayed in the military, please rate the following on a scale of 1 to 10 with 1 being not important and 10 being very important." Forty-three percent of military spouses said that the availability of career opportunities for both spouses was a very important factor, providing evidence of a potential direct link between stressors related to employment and service member retention intentions.[20]

---

[19] The "Impact of Moving" measure was based on eight attitudinal questions: (1) Moving has had a positive impact on my family, (2) We move more frequently than I would like, (3) Moving has provided me with many positive opportunities, (4) Moving has allowed me to make new friends, (5) We have moved to exciting places, (6) Moving is difficult on our children, (7) One of the benefits of being a military spouse is getting to move, and (8) I like to move. These were aggregated to a five-point scale.

[20] "Very important" was defined as a ranking between 8 and 10 on a 1- to 10-point scale, with 10 being the most important.

There is also evidence that military spouses' support to stay on active duty and satisfaction with military way of life are negatively correlated with worsening financial conditions, depression or anxiety, and stress (DMDC Military Family Life Project, 2015). Additionally, spousal satisfaction with military life is also negatively correlated with problematic behaviors among military children. Literature also exists demonstrating that PCS moves are correlated with increasing spousal stress (Drummet, Coleman, and Cable, 2003; Morgan, 1991) and child stress and disruptions (Clever and Segal, 2013; Drummet, Coleman, and Cable, 2003). As such, there is suggestive evidence that frequent relocation can add to financial stress (as reported earlier) and other forms of stress and that this has the potential to impact retention outcomes.

In addition, many studies report that spouse preferences influence service member preferences in general, which suggests that spouse preferences also influence service member retention intentions (Bourg and Segal, 1999; Krauss, 1996; Lakhani, 1995; Lakhani and Fugita, 1993; Mohr, Holzbach, and Morrison, 1981; Meadows, Tanielian, and Karney, 2016; Orthner, 1990; Rosen and Durand, 1995). In general, this set of studies is quite dated, suggesting a need for a new study with more-current data. The most recent correlations estimated between spouse and service member preferences come from the RAND DLS Study (Hengstebeck et al., 2016). That study found positive correlations between spouse and service member preferences for retention intentions, with a correlation reported at 0.43. This correlation does not account for any demographic characteristics (e.g., age, gender, minority status), service member characteristics (e.g., service branch, pay grade), or spouse socioeconomic status (e.g., employment, educational attainment).

Taken together, these studies suggest that PCS moves could indirectly affect service member retention intentions by impacting spouse satisfaction through several different channels, including spouse employment, stress, and other factors.

While the literature provides evidence of a link between spousal employment difficulties and spousal retention intentions, we were not able to find evidence linking spousal employment difficulties to actual retention outcomes. A 2012 study by Hosek, Asch, and Mattock may provide an explanation. In 2009, regular military compensation, which includes basic pay, allowances for subsistence and housing, and the federal tax advantage deriving from the fact that allowances are not taxed, was above the 80th percentile of comparable civilian earnings. Applying this statistic to the average service member total pay figure of $55,367 in Burke and Miller (2017), we find that the average service member was paid about $20,763 more than the average comparable civilian. This premium more than covers the estimated $3,100 reduction in spousal earnings caused by a PCS move (Burke and Miller, 2017), which suggests that remaining in the military may be the better choice for the family as a whole, at least from a financial standpoint. We acknowledge that this argument does not account for the social and psychological impact of being underemployed or unemployed, which may also influence the family's retention decision.

Changes in Retention-Related Outcomes over the PCS Cycle

Complementing the findings from the existing literature about possible disruptions to spousal employment and retention intentions following a PCS move, we use the DLS data to estimate how retention-related outcomes for service members, spouses, and teenaged children change before and after a PCS move.[21] In particular, we use statistical analysis to describe outcomes prior to and after a PCS move for multiple respondent types in the family (i.e., service member, spouse, and teenaged child), including short- and long-term periods before and after a move. The full paths of these outcomes over time are called "trajectories"[22] and allow one to see how one or more outcomes of interest change over a specified time period. The outcomes discussed in this section are factors that may influence a service member's and his or her family's decision to continue with service or leave the military. They include financial well-being, commitment to the military, and satisfaction with military life, as well as retention intention itself. More details about the DLS and the methods used in the analysis are presented in Appendix B. It is important to note that the DLS is not representative of all military families, since it includes data from only those that are married (to include families with and without children). Additional restrictions during analysis (e.g., including only married families that experience a PCS move) further limits generalizability.

Table 2.2 lists the specific outcomes that could be related to PCS moves, investigated for service members, spouses, and teen children. For the financial stress scale, a higher number indicates more stress or a worse outcome. For the other outcome variables, commitment, satisfaction, and retention intentions, a higher number indicates a better outcome in terms of more-favorable attitudes and preferences. Both service members and spouses provided separate reports of their outcomes and moves. Teens were children of service members and spouses ages 11 and over who were invited to participate as a separate respondent. Teens provided separate reports about their own outcomes, and their PCS move information was based on the spouse (or parent) report.

We conducted two statistical tests to determine whether the trajectory of the outcomes was different over the course of the PCS move. First, we tested to see whether the overall trajectory was different from a flat line (i.e., no changes in average values over the move cycle). Second, we tested to see whether the overall trend differed significantly from a straight line (i.e., a constant change over time that is unrelated to the phase of, or timing within, the PCS move period). Our results indicate that service member commitment to the military, service member satisfaction with military life, and spouse financial stress around a PCS move are significantly different from both a

[21] This analysis is descriptive, as we cannot make any causal claims about the effect of PCS moves on outcomes. In addition, we do not know when service members received notice that they would be relocating, meaning that some of the service members might have known a move was imminent, while others might not have.

[22] Unlike the other quantitative results in this report, the analysis of trajectories prior to and after a PCS move has not been published elsewhere.

Table 2.2
Deployment Life Study Trajectory Analysis Outcome Variables

Outcome	Family Members	Range
Financial stress scale[a]	Service member, spouse	(lowest) 1 to 4 (highest)
Commitment to military[b]	Service member, spouse, teen	(lowest) 1 to 5 (highest)
Satisfaction with military life[c]	Service member, spouse	(lowest) 1 to 5 (highest)
Retention intentions[d]	Service member, spouse, teen	(lowest) 1 to 5 (highest)

[a] Includes four items assessing current financial condition, difficulty paying bills, ability to save money, and concern about current financial situation.

[b] Includes three items assessing the degree to which being part of the military inspires the respondent to do the best they can as a service member/spouse/military child, the degree to which the respondent is willing to make sacrifices to be in the military, and the degree to which the respondent is glad they/their spouse/their parent is part of the military.

[c] For service members: Generally, on a day-to-day basis, how satisfied are you with the military way of life? For spouses: Generally, on a day-to-day basis, how satisfied are you with the quality of your life as the spouse of a service member?

[d] For service members: Assuming you have a choice to stay on active duty or not, how likely is it that you would choose to stay on active duty? For spouses: Which of the following statements describes how you feel about whether your spouse should stay on or leave the military? (Strongly favor staying [5] to strongly favor leaving [1]). For teens: Do you think your military parent should stay on or leave the military beyond [his/her] service obligation? (Strongly favor staying [5] to strongly favor leaving [1]).

flat and a straight line, suggesting that these outcomes change during the move cycle and that these changes are not constant within the move cycle. Service member retention intentions are marginally statistically significant.[23] None of the other outcomes we examined provided statistically significant results, so they are not reported here (see Appendix B for more details).

These tests do not tell us anything about the direction of the trends. For that, we reviewed plots of the predicted outcomes over time. In Figures 2.6 through 2.9, the *x* axis represents time, with vertical black lines demarking the phases of the PCS period, and the *y* axis depicts the predicted average outcome, based on the 5-point survey scale. This axis is centered on the mean value and presents a range representing plus or minus one standard deviation from the mean. In general, these plots serve as descriptive tools for understanding how service members and their families changed over the course of a PCS move. The trajectory plots allow for examination of both short- and longer-term outcomes, both before and after a PCS move. The plots show predicted values for those outcomes over a hypothetical 12-month period, with a PCS move occurring at the midpoint. The two-month window before and after the move

[23] Service member retention intention trajectory was statistically significantly different from a flat line but just missed the established threshold for statistical significance for the test for whether it was different from a straight line. Because it almost reached statistical significance and tells a similar substantive story to the other outcomes explored, we have chosen to include it here.

reveals more-immediate changes, whereas the six-month window gives us some indication of baseline functioning before a move and longer-term functioning post-move.[24]

It is important to note that because the statistical tests we ran tested the overall pattern of the trajectories, all we can say with statistical certainty is that these patterns overall differ from both a flat and straight line. We cannot speak to individual significant changes in levels or slopes at particular points within the trajectories (except for cases where these are explicitly tested).

For service members, commitment to military life shows a U shape during the move period. In the longer run prior to the PCS move, service members show a (slightly) increasing slope (see Figure 2.6). The slope declines in the two months prior to the PCS, and then the slope increases again in the two months after the PCS move, followed by a slight decline in the longer term after the move. The U-shaped pattern suggests that, while there is a dip in commitment during the move, commitment increases again after the move period. This pattern is consistent with anecdotes from service members who say that notification of a move and, especially, the actual preparation for a move are dissatisfying aspects of military life, which could lead to a decrease

Figure 2.6
Service Member Commitment to Military over the PCS Move

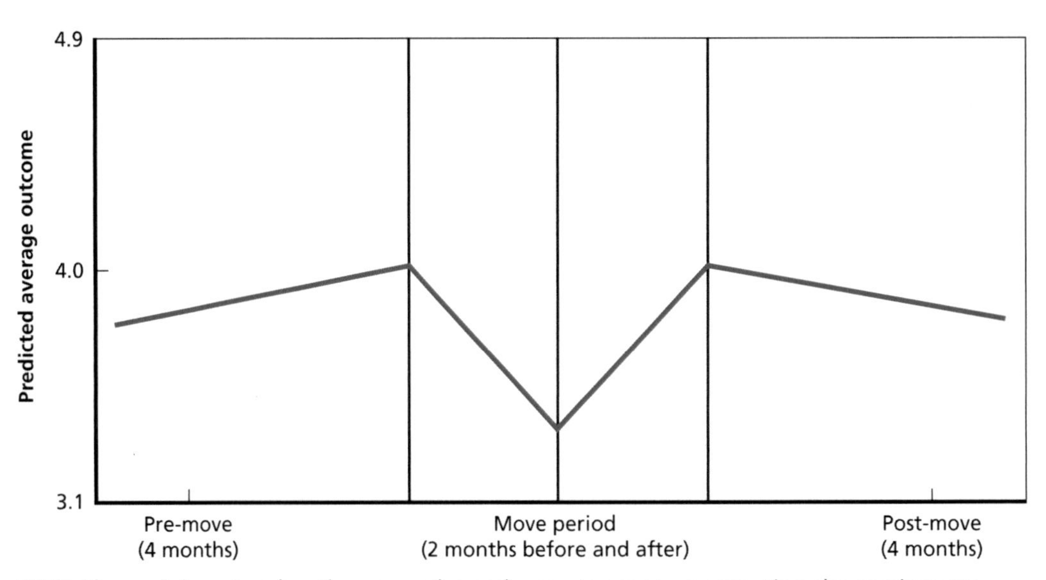

NOTE: The *y* axis is centered on the mean value and presents a range representing plus or minus one standard deviation from the mean.

RAND *RR2304-2.6*

---

[24] Our decision to use six months as the longer-term time period was based on information from the Navy. In data provided to the sponsor, the Navy indicated that the Chief of Naval Operations has established a goal of an average of six months of lead time for sailor notification of an upcoming PCS.

in overall commitment to the military. However, after the move has taken place and life returns to some state of normalcy or routine, commitment once again increases.

We see a very similar but less pronounced pattern for retention intentions as we do for commitment to the military (see Figure 2.7). Within the two-month window prior to a move, there is a steep decline in intention to remain in the military; however, it rises again in the two months after the move and remains relatively stable for the rest of the observation period. Note that this trajectory is only marginally significantly different from a constant line, so the less pronounced trend is not surprising.

For service members, reported satisfaction with military life shows a relatively flat trajectory *except* during the two months immediately before a move, where we see a steep decline in reported satisfaction with military life (see Figure 2.8). It is worth noting that we also see a statistically significant increase in satisfaction right at the time of the PCS move, and reported satisfaction is relatively stable from there on. That suggests that military satisfaction returns to a stable value more quickly than do commitment and retention intentions.

For spouses, financial stress is relatively stable in the longer run both before and after the PCS move period (see Figure 2.9). However, financial stress increases (positive slope) during the short-term move period (i.e., two months before and after the PCS move). The slight drop at the time of the move is not statistically significant.

The DLS analysis demonstrates that most of the change (i.e., decline) in the predicted average trajectories of service member satisfaction, commitment, and retention

Figure 2.7
Service Member Retention Intentions over the PCS Move

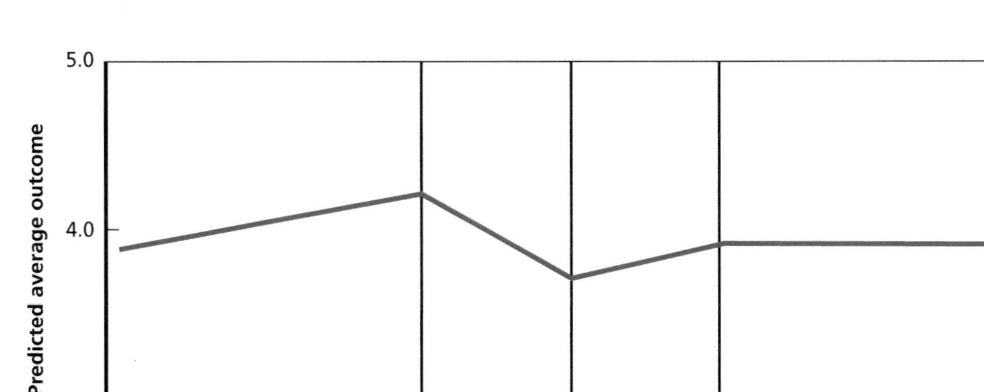

NOTE: The *y* axis is centered on the mean value and presents a range representing plus or minus one standard deviation from the mean.

RAND *RR2304-2.7*

Figure 2.8
Service Member Satisfaction with Military over the PCS Move

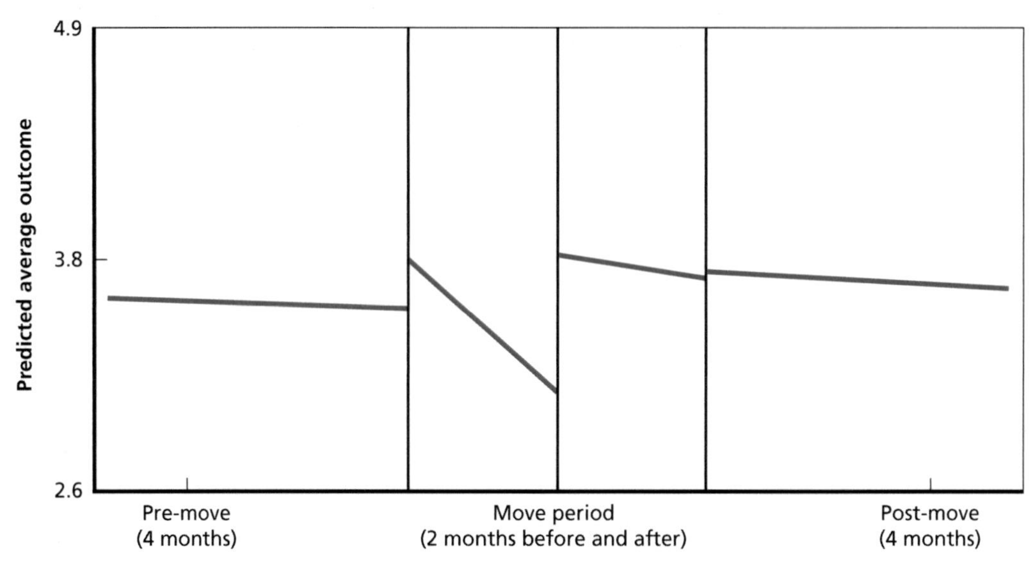

NOTE: The *y* axis is centered on the mean value and presents a range representing plus or minus one standard deviation from the mean. The model includes steps at –2 months, 0 months, and +2 months from the PCS move. For further details, see the discussion on page 63.
RAND *RR2304-2.8*

Figure 2.9
Spouse Financial Stress over the PCS Move

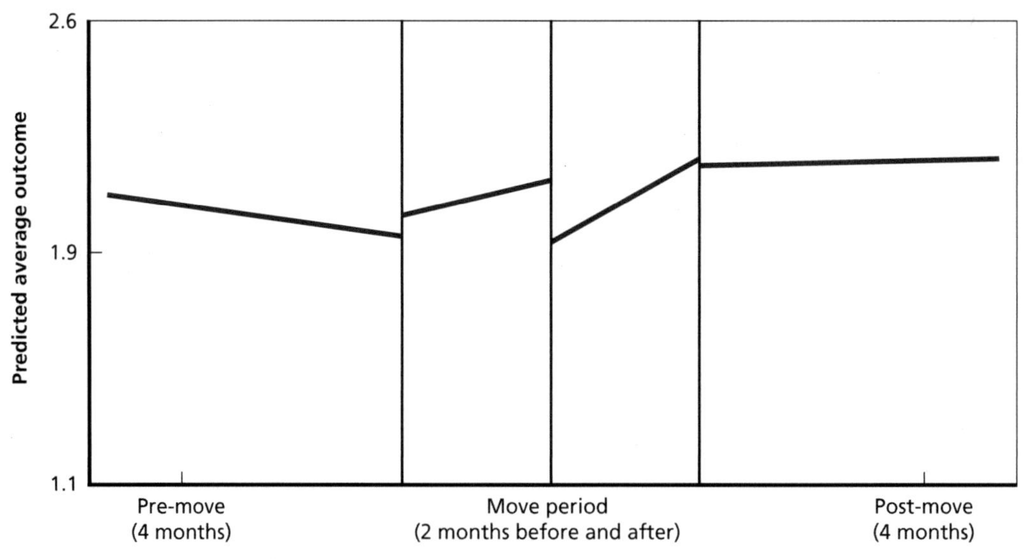

NOTE: The *y* axis is centered on the mean value and presents a range representing plus or minus one standard deviation from the mean. The model includes steps at –2 months, 0 months, and +2 months from the PCS move. For further details, see the discussion on page 63.
RAND *RR2304-2.9*

intentions occurs in the short term—right around the time of the move. Service members' reported military commitment and retention intentions show a dip right at the time of the move but then increase shortly thereafter. Satisfaction with the military drops during the two months immediately before a move, but, again, this drop is only temporary and begins to rise right around the time of the PCS move. The results also provide evidence that spouse financial stress increases during the short-term move period (defined as the two-month window around a PCS move) and that the changes tend to be temporary, on average.

Timing of PCS Moves and Deployments

We also reviewed the literature about the possible interaction between PCS moves and deployments, their prevalence, and the potential effect they may have on family stability if they co-occur within a short period of time. Although PCS moves and deployments are not equivalent experiences, the associations between PCS moves and retention intentions that we discussed above could be exacerbated if deployments and PCS moves are related to common problems that have been linked to service member retention intentions, especially if these two distinct events occur within a short time frame.[25] As a result, some of the adverse effects that occur around a PCS move could be partially attributed to a recent deployment and might not be a direct result of the PCS move itself.

The existing literature demonstrates that PCS moves and deployments generate similar disruptions to family stability (Angrist and Johnson, 2000; Chandra, Martin, et al., 2010; Chandra, Lara-Cinisomo, et al., 2010; Chartrand et al., 2008; Clever and Segal, 2013; DMDC Military Family Life Project, 2015; Drummet, Coleman, and Cable, 2003; Hosek and Wadsworth, 2013; Lester et al., 2016; Morgan, 1991; Richardson et al., 2011; SteelFisher, Zaslavsky, and Blendon, 2008; U.S. Chamber of Commerce, 2017). The associations between PCS moves and retention intentions discussed earlier might potentially be overstated, given that deployments and PCS moves are related to common problems that have been linked to service member retention intentions, either directly or indirectly. In particular, evidence suggests that the length of a deployment is associated with reductions in spouse employment, and, as discussed earlier, PCS moves have been found to reduce spouse employment (Angrist and Johnson, 2000; SteelFisher, Zaslavsky, and Blendon, 2008). If PCS moves and deployments co-occur in a short period of time, problems related to spousal employment could increase. Although the definition of a short period of time is subjective, we define this in Table 2.3 as less than 15 months and further describe the frequency

---

[25] To our knowledge, the existing literature does not contain recent representative information about the timing of PCS moves relative to deployments.

Table 2.3
Distribution of Service Members and Spouses with a PCS Move by Timing Between the PCS Move and Deployment

Months Between PCS Move and Deployment	Service Member	Spouse
≤ 3 months	14%	15%
> 3 months, ≤ 6 months	10%	10%
> 6 months, ≤ 9 months	8%	7%
> 9 months, ≤ 12 months	6%	6%
> 12 months, ≤ 15 months	4%	3%
> 15 months	3%	3%
PCS move, no deployment	54%	56%
Number with a PCS move	737	853

NOTE: Unweighted tabulations constructed using DLS data.

of these joint events occurring within even smaller time frames (e.g., three and six months).

In addition, deployments have been linked with spouse stress, spouse mental health, family functioning, and financial strain, as well as child behavioral outcomes and child stress (Chandra, Martin, et al., 2010; Chandra, Lara-Cinisomo, et al., 2010; Chartrand et al., 2008; DMDC Military Family Life Project, 2015; Drummet, Coleman, and Cable, 2003; Hosek and Wadsworth, 2013; Lester et al., 2016; Morgan, 1991; Richardson et al., 2011; U.S. Chamber of Commerce, 2017). Above, we documented that spouse satisfaction is negatively correlated with stress, anxiety, and adverse child outcomes. Previous research demonstrates that repeated exposure to family stress results in a "pile-up of demand," meaning that the effects from multiple stressful situations accumulate over time. This concept has been applied to military families to show that previous experiences can impact the amount of strain caused by a new "crisis" or stressful event, including relocation (e.g., Lavee, McCubbin, and Patterson, 1985; McCubbin and Patterson, 1983). As a result, if deployments and PCS moves occur close together, then this negative correlation with spouse satisfaction may be amplified.

To get a sense of the scope of the problem, we used the DLS data to look at the distribution of service members and spouses who experienced a deployment and a PCS move within specific time intervals. As mentioned earlier, these data are not representative of the entire military population. However, for the DLS sample, we can construct these statistics to get a sense of how many married families who experienced a PCS move also experienced a deployment around the same time. Table 2.3 contains our tabulations of the distribution of service members and spouses with a PCS move by timing between the PCS move and deployment. These statistics are restricted to those who had a PCS move during their three-year survey period. For those who had more than one PCS move and deployment pair, we counted the pair with the smallest time interval.

The second column depicts the distribution of service member respondents with a PCS move by timing between the move and deployment. The third column depicts the distribution of military spouse respondents with a PCS move by timing between the move and deployment.[26] We found that about 15 percent had a deployment and PCS move within three months of each other and about 25 percent experienced these two events within six months of each other. Although the quantitative effects of short time periods between PCS moves and deployments are unknown at this time, these data show that while more than half of married families who experienced a PCS move were not deployed over the three-year study period, nearly a quarter experienced a PCS move and deployment within six months of each other. This latter group could be experiencing a greater level of family disruption, on average.

Timeliness of Receiving PCS Orders

The timeliness of PCS orders and the frequency with which they change also have the potential to exacerbate family disruptions and, therefore, influence retention intentions.[27] Data from the DMDC Status of Forces Survey of Active Duty Members suggest that a change in PCS orders was not a problem for the majority of service members. As Figure 2.10 shows, 62 percent of service members reported that a change to PCS orders was not a problem, while 26 percent of service members reported at least a moderate problem. Compared with changes in PCS orders, time to prepare for a move was a somewhat bigger problem. As Figure 2.11 shows, 49 percent reported that time to prepare for a move was not a problem, while 34 percent reported that it was at least a moderate problem. Figure 2.12 shows 2016 data on the extent to which receiving orders in a timely manner was a problem. The figure shows that 52 percent reported that PCS order timeliness was not a problem, while 35 percent reported at least a moderate problem.

To get a sense of whether there have been changes over time in the extent to which a change in PCS orders or time to prepare for a move was reported as being a problem, we present data spanning 2003 to 2016 in Table 2.4.[28] The percentages of respondents reporting that a change in PCS orders was not a problem, was a small or moderate problem, and was a large or very large problem were stable over this time

[26] The counts of service members and spouses experiencing a PCS move are not equal due to missing data. It is also possible that families decided that not all family members would move or that not all family members would move at the same time (e.g., service member moves first).

[27] To our knowledge, the existing literature does not discuss the direct impact of timeliness of PCS orders on family disruptions or retention intentions.

[28] Historical information about the extent to which receiving PCS orders in a timely manner was a problem is not available.

Figure 2.10
For Most Recent PCS Move, to What Extent Was a Change in PCS Orders a Problem?

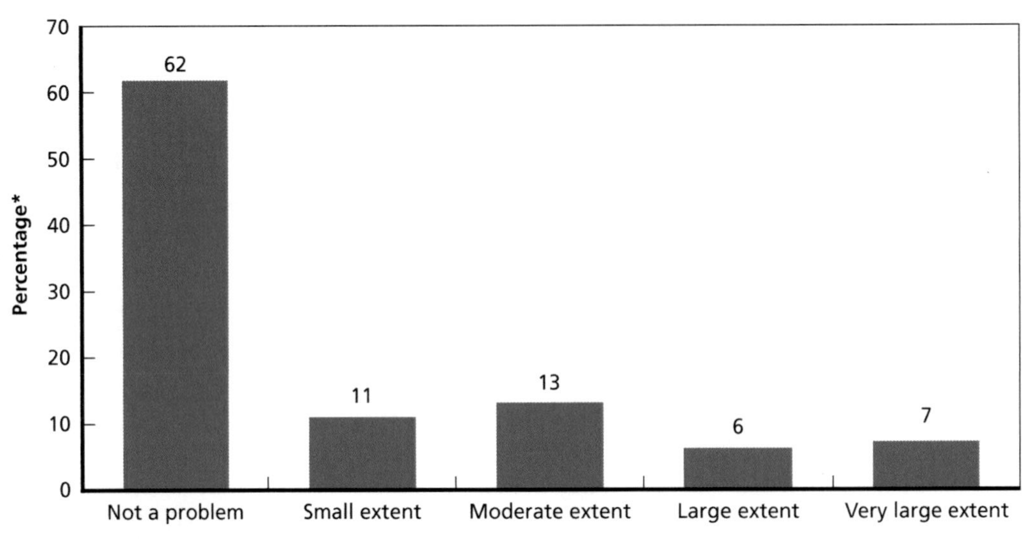

SOURCE: DMDC, 2017.
* Percentage of respondents who answered the relevant survey item and had experienced at least one PCS move at any point in the past.
RAND *RR2304-2.10*

Figure 2.11
For Most Recent PCS Move, to What Extent Was the Amount of Time to Prepare for a Move a Problem?

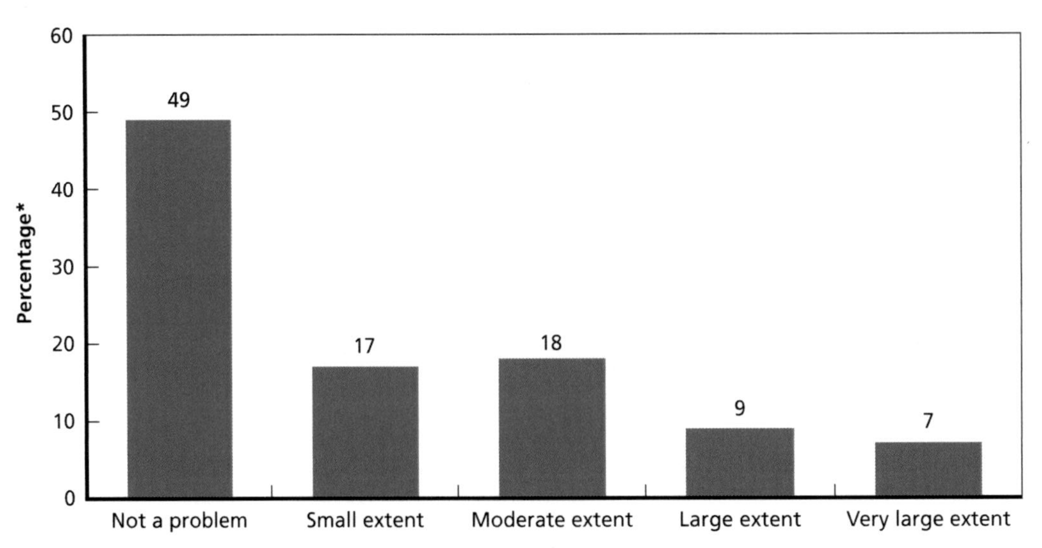

SOURCE: DMDC, 2017.
* Percentage of respondents who answered the relevant survey item and had experienced at least one PCS move at any point in the past.
RAND *RR2304-2.11*

Figure 2.12
For Most Recent PCS Move, to What Extent Was Receiving Orders in a Timely Manner a Problem?

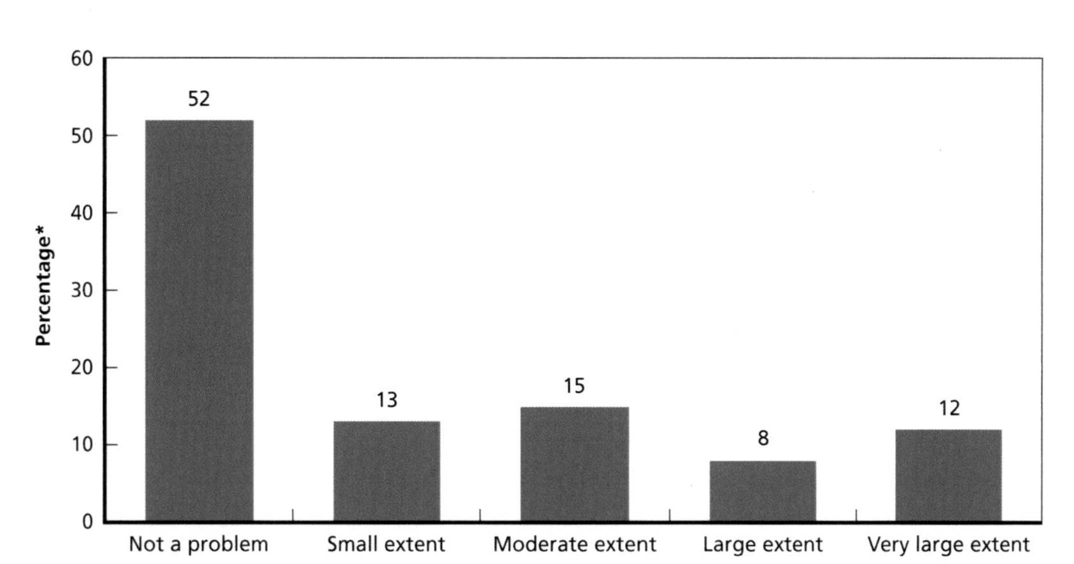

SOURCE: DMDC, 2017.
* Percentage of respondents who answered the relevant survey item and had experienced at least one PCS move at any point in the past.
RAND *RR2304-2.12*

period. In addition, there was a small increase in the percentage reporting that the time to prepare for a move was not a problem between 2003 and 2016, once the margins of error are taken into account.

We note that these data rely on self-reported information, and what is considered a problem varies across service members. Another limitation of these data is that these are constructed using repeated cross-sections, meaning that these tabulations represent a particular point in time and these surveys do not follow the same people over time. Furthermore, with respect to the question regarding receiving a PCS order in a timely manner, the interpretation of when orders are received and what is considered timely is also different across service members.

Analysis of our interview results with military personnel experts provides additional insight into the types of issues related to timeliness that service members face. The most frequently mentioned timeliness issue among interviewees was moving during summer months, or "peak periods." While moving during summer months reduces disruptions related to children changing schools, it can exacerbate disruptions related to household management, demonstrating that addressing one set of PCS move disruptions could adversely impact another type of disruptions. To move household goods, the military uses contractors, and these contractors become scarce in summer months, when nonmilitary families are also likely to move. Earlier bookings could

Table 2.4
Trends in PCS Order–Related Problems, 2003 to 2016

For Your Most Recent PCS Move, to What Extent Was the Following a Problem?	2003	2004	2007	2013	2016
Change in PCS orders					
Not a problem	61%	64%	63%	65%	62%
Small extent/moderate extent	24%	24%	23%	21%	24%
Large extent/very large extent	15%	12%	14%	13%	13%
Margin of error	+/– 2%	+/– 1 to 2%	+/– 1 to 2%	+/– 2%	+/– 3%
Amount of time to prepare for a move					
Not a problem	40%	44%	45%	52%	49%
Small extent/moderate extent	39%	38%	36%	32%	35%
Large extent/very large extent	21%	18%	19%	17%	16%
Margin of error	+/– 2%	+/– 1 to 2%	+/– 1 to 2%	+/– 2%	+/– 3%
Receiving orders in a timely manner					
Not a problem			Not available		52%
Small extent/moderate extent					28%
Large extent/very large extent					20%
Margin of error					+/– 3%

SOURCES: DMDC, 2004; DMDC, 2006; DMDC, 2008; DMDC, 2014; DMDC, 2017.

NOTE: Percentages are limited to respondents who answered the question and had experienced at least one PCS move at any point in the past.

increase the options available to an individual household at a given move time if there is a positive correlation between lead time and availability of contractors.

The second–most frequently mentioned timeliness issue was inadequate notice of upcoming PCS moves, which reinforces the statistics presented in Figures 2.11 and 2.12 showing that more than one-third of respondents reported that the amount of time to prepare for a move or receiving orders in a timely manner was at least a moderate problem. The ability to initiate the planning process prior to receipt of funded orders may help alleviate this stress.

When we asked interviewees about issues related to the PCS assignment process, the two most frequently discussed issues were (1) movers, the moving process, and moving logistics and (2) the U.S. Coast Guard (USCG) approach to PCS moves, which standardizes move dates by rank and aims to relocate service members during summer months. Problems with logistics of the move and the USCG approach further highlight issues related to moving during peak periods, when resources to conduct these moves are likely scarce. Coast Guard interviewees offered both positives and negatives to focusing PCS moves in the summer months, as this conversation with three different individuals reveals:

> *Interviewee 1:* I think the move process is still shaky. A member puts requests out for move and people put out bids [moving companies]. If no one bids you won't get the move. It's a shaky system of contractors and subcontractors. A company can pick up a bid and not show up if they get a better offer.
>
> *Interviewee 2:* The number of moves that happen in the summer time puts a big strain on the system. You don't know how things will line up and whether there are enough movers. The demand is cyclical.
>
> *Interviewee 3:* I think the Coast Guard actually feels it [problems associated with moves] because we only move in the summer. But if you move throughout the year you get additional problems with getting to places in time for the school year.

Our interviews with SMEs suggest that the availability of funding for PCS moves may influence timing of notification. We learned from time-series data provided by the Navy that lags in congressional funding are directly associated with when orders are issued. One interviewee stated, "I think the biggest issue that was brought up was how the funding comes from Congress that allows us to write orders in a timely manner, and that funding comes in fits and starts, which causes all kinds of bad things on the back side. It starts with how we get the money and how we're allowed to spend it." If a service member cannot begin the planning process for a move until required funding has been cleared, then any delays in appropriations may add to the stress associated with a move, increased logistical problems (e.g., finding movers, enrolling children in a new school, finding suitable housing), and ultimately reduced satisfaction and retention intentions.

Finally, timing of PCS orders may also be related to a technical issue that, if resolved, could potentially ease the PCS process. During our interviews with SMEs who work on the logistical aspects of PCS moves, we learned that, currently, the assignment system (which triggers a PCS move) is not linked with other systems that address PCS logistics (e.g., hiring movers). This lack of automation means that service members must understand the entire PCS process and know not only the order of steps to take but also who to go to in order to take care of those steps. Interviewees suggested that many existing websites are not user friendly. One interviewee proffered that an easy-to-use process, in which the PCS assignment and logistics systems are electronically linked and able to help with coordination efforts, might alleviate some of the (real or perceived) negative aspects of moves. Another respondent, from a logistics-related agency, stated,

> We need to get the orders earlier so that service members can plan. We need to have linkages so that we receive a service member's orders electronically and automatically in the system and then can talk directly to the service member about their specific needs. But it needs to happen electronically so we know who needs to go where and when. Then the earlier they answer these simple questions about the property they need to move, the better we can help them.

During this same discussion, the Defense Travel Management Office (DTMO) was offered as the "ideal" for a consistent, DoD-wide model for how to handle the actual logistics surrounding a PCS move. DTMO addresses all travel needs for all service members, across the service branches, within a single office. This consistency was seen as a facilitator by interviewees, who suggested that without it, the service branches are "doing their own thing." One interviewee even went so far as to recommend that the actual management of moving service members' goods be shifted out of the U.S. Transportation Command (TRANSCOM) and into OUSD(P&R), which, it should be noted, also owns DTMO. As one interview noted, "TRANSCOM moves bedroom furniture . . . but should it? The point is that their focus is getting the Army to the fight, not getting the Army spouse's china cabinet to Germany." We caution, however, that shifting responsibilities for logistical aspects of PCS moves to OUSD(P&R) is but one possibility for integrating relocation services across DoD. It was beyond the scope of this study to complete a full cost-benefit analysis of such a recommendation.

Moving as a Family Unit Versus Separately

The extent to which the service member and his or her family members move at the same time versus at different times (separately) is understudied. To our knowledge, there is no data source or evidence available on the size of the population that moves as a family unit versus at different times. The interviews with experts from each service branch suggest that the branches do not collect data on the frequency with which each type of move occurs. While an extensive literature shows that school changes and parental absences typically worsen outcomes for children,[29] it is not clear how moving as a family unit versus moving separately at different times affects family stability. The impact likely depends on a family's circumstances (e.g., whether separation allows children to finish a school year, whether the spouse works or is in school). Several studies provide some information about the potential reasons why families might decide to

[29] Studies find that changing schools worsens test scores and that high turnover negatively impacts movers and non-movers within schools (Hanushek, Kain, and Rivkin, 2004; Rockoff and Lockwood, 2010; Schwartz et al., 2011; Schwartz et al., 2017; Schwerdt and West, 2013). Studies also find that parental absences identified by variation in deployments generally worsen child outcomes, including student test scores, school attendance, homework completion, and behavior problems (Chandra, Martin, et al., 2010; Chandra, Lara-Cinisomo, et al., 2010; Chartrand et al., 2008; Engel, Gallagher, and Lyle, 2010; Lyle, 2006; Richardson et al., 2011).

live separately, some of which are related to family disruptions discussed earlier in the chapter.

Maury and Stone's (2014) survey asked spouses about living separately from their active-duty spouses. As mentioned, that study used a convenience sample, and its findings are not necessarily representative of the military spouse population. Sixty percent of military wives reported having resided in a different geographical location than the service member, which includes separation due to deployment. Among those having ever experienced a separation, 18 percent reported maintaining a career after a PCS move, and 34 percent reported deployment/family separation as a reason for living separately, with the reasons not necessarily being mutually exclusive. Furthermore, that study does not provide information on how long the separations were on average or whether these separations were temporary or permanent.

A 2001 GAO report provides statistics on the prevalence of instances when the spouse and family were not with the service member at his or her permanent duty station, including both assignments overseas and within the continental United States (CONUS). Fourteen percent of military personnel with families were not accompanied by their families, with this percentage varying by marital and dependent child status. Among unmarried personnel with dependent children, 51 percent were unaccompanied. Among married personnel, 16 percent without dependent children and 9 percent with dependent children were unaccompanied. The top reasons for family members not accompanying a service member on his or her current tour, among those who were married and/or had dependent children, were separation or divorce (30 percent), spouse's career (22 percent), and children's education (12 percent). Because GAO used data from 1999, these statistics may not necessarily be an accurate depiction of the current prevalence of the separation of families, but they at least suggest that reasons in the past for a family separation include spouse employment and children's schooling.

Summary

In this chapter, we presented a broad list of family disruptions generated by PCS moves, demonstrating that these disruptions have the potential to impact all members of a military family. We focused on a subset of disruptions to study further, including spouse employment, service member retention intentions, and disruptions related to moving as a family unit versus separately. While we found ample evidence that PCS moves disrupt spouse employment and retention intentions, there was no evidence available on the extent to which families typically move as one unit versus sequentially or on whether either type of move mitigates or exacerbates PCS-related disruptions. With the exception of the study by Burke and Miller (2017) that provides causal estimations of the effect of PCS moves on spouse employment, we note that much of the research cited present correlational relationships of the impact of PCS moves on spouse employment and retention intentions. This correlational evidence suggests that PCS moves can contribute to undesirable spouse employment outcomes and could generate difficulty finding employment. The literature provides correlational relationships between PCS moves and service member retention intentions through direct and indirect channels. Analysis of DLS data further provides evidence that retention intentions themselves and factors that could impact retention intentions (e.g., service member satisfaction and commitment, spouse financial stress) are temporarily, negatively correlated with PCS moves. These negative associations between PCS moves and retention intentions may be amplified if deployments occur around the same time as PCS moves because, as the literature demonstrates, both events are correlated with similar stressors that have been associated with retention intentions. Further, issues with timeliness of PCS orders were presented using published DMDC tabulations and information obtained through interviews with SMEs, including adverse effects in the timing of notification due to uncertain congressional funding and technical problems.

Our review is based on preexisting data sources and thus reflects the interests and representativeness of the original studies. Future research, especially involving primary data collection about the impact of PCS moves on specialized populations (e.g., more-junior pay grades or dual military couples) and topics of interest, could help fill any potential gaps in current understanding.

In the next chapter, we discuss existing programs and policies that address family disruptions associated with PCS moves to determine whether there are gaps that need to be filled.

CHAPTER THREE

Existing Programs Addressing Disruptions

DoD and the service branches offer a range of policies, programs, and services to assist service members and families with the disruptions that occur as a result of PCS moves (as well as other aspects of military service). This chapter details the results of RAND's research into existing programs and whether these programs address all of the identified disruptions or whether gaps exist. The list of existing policies, programs, and services is not necessarily exhaustive but is meant to demonstrate the scope of what is currently available to service members and their families. The chapter begins by listing existing DoD and service branch–specific programs and presents a crosswalk to the first- and second-order disruptions described in Chapter Two. We also discuss assignment policies related to PCS moves and their relevance to family stability. More details about DoD and service branch–specific programs are provided in Appendix C. Appendix D provides (self-reported) service branch–specific information on assignment policies related to PCS moves.

While there are many programs and services offered by DoD, it is worth mentioning that there are other organizations that offer assistance to military families during PCS moves, some of which overlap with existing DoD programs and service branches. For example, there are nonprofit organizations that offer counseling, team mentoring, and emergency financial assistance to help cover PCS costs. This could suggest redundancies with DoD programs or could simply be a result of the existence of stakeholders outside of DoD that are interested in helping military families. These non-DoD organizations are excluded from our study.

DoD Programs Addressing PCS Disruptions

A list of programs and services offered by DoD that serve the entire military is provided in Table 3.1. These military-wide programs and services provide assistance with moving logistics, counseling, spouse education and employment, and child-related issues. In addition, DoD has a centralized website, Military OneSource, that contains information about military programs and services offered.

Table 3.1
Programs and Services Offered by DoD

Programs and Services	Purpose
Military OneSource	Centralized website with information about military programs and services
Housing Early Assistance Tool (HEAT) MilitaryINSTALLATIONS Transportation Office (TO) Move.mil Plan My Move	Programs to assist with moving logistics
Military Family Life Counseling Program	Nonmedical counseling and briefings offered to service members and families
Spouse Education and Career Opportunities (SECO) Military Spouse Employment Partnership DoD State Liaison Office Initiatives	Programs that provide military spouses with a broad range of resources to facilitate the pursuit of education and employment
MilitaryChildCare.com Military Kids Connect	Resources to assist with child care and to support children

NOTE: Programs and services listed are not exhaustive but are meant to be illustrative of the types of programs and services available.

Service Branch–Specific Programs Addressing PCS Disruptions

In addition to DoD-wide programs, a broad range of existing programs addressing PCS move disruptions are service branch–specific. DoD Instruction 1342.22, which was issued in 2012, updated requirements related to the provision of family readiness services to military families. Service branches are required to provide a broad range of programs to promote family readiness covering relocation assistance and spouse education and career opportunities. The access points for these services include installation-based military and family readiness centers and online resources, such as Military OneSource.

Each service branch has its own versions of military and family readiness centers that are community-based and installation-specific. These centers offer similar types of services. These core programs are described as follows:

1. *Child and youth programs:* Each installation offers its own version of a child and youth program. In general, these programs offer various types of child care, including child development centers, programs for school-aged children, and family child care homes. These programs also typically include recreational activities and support to help children assimilate to a new installation. Depending on the installation, the child and youth program may include partnerships with various organizations, such as the Boys & Girls Club of America and

the 4-H club. Each installation also has a school liaison officer who serves as the main contact person on school-related issues unique to military families, including problems related to frequent moves and deployments.

2. *Employment assistance:* Each installation's military and family readiness center offers a program to help military spouses find employment after a PCS move. Employment assistance includes help with resume building, application assistance, career planning and counseling, information about job and volunteer opportunities, and interview techniques. In addition, these programs typically offer various classes and workshops and individualized services. Depending on the installation, some of these employment assistance program services are delivered in person, over the phone, or via email. Often these programs also cover financial management and educational opportunities. These programs also provide information on the military-wide SECO program.
3. *Exceptional Family Member Program:* The EFMP provides services to military families that include a member with special needs (e.g., physical, developmental, mental health disorders). The EFMP provides community, housing, educational, medical, and personnel services support through coordinated efforts between military and civilian agencies.
4. *Financial counseling:* Each installation's military and family readiness center offers financial counseling that covers such topics as budgeting, managing debt, and retirement planning. These services also cover financial planning during life events, such as PCS moves, home purchases, and saving for children's college education.
5. *Readiness:* Each installation's military and family readiness center offers services to promote readiness. These services are offered as workshops, trainings, and tutorials both online and in person and cover such topics as relocation, deployments, separations, and family functioning.
6. *Relocation assistance:* Each installation's relocation assistance program generally covers how to plan a move, issues specific to foreign-born spouses, referrals to emergency financial assistance, additional assistance for first-time movers, and relocation sponsors.
7. *Unit-sponsored volunteer group:* Each service branch has its own version of a command- and unit-sponsored volunteer group in which military spouses typically play a key role. The goals of these programs generally include maintaining the flow of information between military families and command; providing families with support, education, and community resources to promote resilience and readiness; and providing activities and support that enhance integration into the military community. In the Air Force, this program is known as the Key Spouse Program. In the Army, this program is known as the Family Readiness Group. In the Marine Corps, this program is known as the Unit, Personal, and Family Readiness Program. In the Navy, each installation has a

command ombudsman who serves as the conduit between the commanding officer and families and a Family Readiness Group, which is a volunteer organization that provides military families with a social network and support group. In the USCG, the ombudsman program is meant to facilitate communication between the command and Coast Guard families.

Our interview analysis shows that the top four frequently mentioned negative aspects of PCS moves were needs related to having an EFMP family member, spouse employment problems, children and school disruption, and family stability and cohesiveness. These issues are addressed by the core set of programs offered by these military and family readiness centers.

Although each service's military and family readiness centers offer the same broad set of programs and services, the specific provisions vary across service branches. Specific provisions were generally obtained directly by the service branches at the request of DoD for input for the congressionally mandated report on family stability and PCS moves. Further information about military and family readiness center programs is needed to determine whether variation across service branches is due to differences in need, available resources, or how much detail was provided by each service branch. Appendix C contains detailed descriptions of additional PCS-related programs offered by each service branch's military and family readiness centers that are not already described in this chapter.

To illustrate the types of programs and services available and how they assist families in mitigating disruptions caused by PCS moves, we provide additional examples below. This list is not necessarily exhaustive but is instead meant to demonstrate the existence of additional programs provided by DoD and USCG:[1]

- The Air Force partners with the nonprofit Air Force Aid Society to offer an orientation program called Heart Link to new active-duty spouses that is held at military and family readiness centers. This program demonstrates scope for coordination and cooperation between DoD and nonprofit organizations to provide services addressing PCS disruptions.
- The Army offers its own version of Military OneSource, called Army OneSource. The program's website includes a tool for booking temporary housing, which may be useful for families who experience a PCS with little lead time.
- The Marine Corps offers family readiness training through its Marine OnLine service, as well as a 24-hour anonymous hotline, called DSTRESS, available to any Marine or family member in need of counseling services. Such services pro-

[1] It is unclear whether the existence of these programs and services indicates differences in need or whether these programs and services are redundant with others in the DoD portfolio.

vide Marines and families with advice and tips for addressing the stress accompanying a PCS move.
- The Navy uses its Child and Youth Program to employ military spouses through a policy that allows military spouse employees to be transferred to the next duty station with the same grade and pay without needing to apply for the position. Roughly half of the program's employees are military spouses. Offering military spouses jobs at military and family readiness centers could assist with providing spouses with portable employment opportunities. However, as mentioned in Chapter Two, Maury and Stone (2014) found that child care was one of the career fields for which military wives had the lowest preference. Therefore, even though employing military spouses at military and family readiness centers assists with continuity of employment among military spouses, these types of jobs may not be the ones that they prefer.
- The Coast Guard offers free and confidential services through its CG SUPRT program, including work-life resources and referrals, nonmedical counseling, health coaching, personal financial wellness, legal services, and education and career counseling. An online library contains other resources, such as videos, self-assessment quizzes, courses, and financial tools.

Existing Programs and Services Appear to Be Comprehensive

To determine the extent to which the first- and second-order disruptions identified in Figure 2.1 are addressed by our compilation of existing programs, we generated crosswalks between them.[2] These crosswalks, depicted in Tables 3.2 and 3.3, demonstrate that DoD provides programs and services that address each of the identified disruptions. As a result, we do not find evidence of any particular gaps. Further, with the exception of family functioning, we find that at least three programs are identified as addressing each of the first- and second-order disruptions identified. Based on our interviews, the most frequently mentioned way to alleviate PCS disruptions was by using existing policies, programs, and services, which, based on our crosswalks, appear to be comprehensive. As we will discuss in Chapter Four, however, future work is necessary to determine the extent to which these existing programs are redundant, whether multiple programs meeting the same needs are desirable (e.g., they target different populations), and ways to improve these programs to better serve military families.

---

[2] These crosswalks do not include the other service branch–specific programs listed above, which target disruptions associated with household management; spouse employment; service member, spouse, and child psychosocial outcomes; military family life; and family functioning.

Table 3.2
Existing Programs and Services That Address First-Order PCS Disruptions

Program	Service Member and Spouse Household Management	Spouse Employment	Child Changing Schools	Family Child Care
HEAT	X			
MilitaryChildCare.com				X
Military Family Life Counseling				
MilitaryINSTALLATIONS	X			
Military Kids Connect				
Military OneSource	X	X	X	X
TO and Move.mil	X			
Plan My Move	X			
SECO		X		
Military and Family Readiness Centers				
Child and youth programs			X	X
Employment assistance		X		
EFMP			X	
Financial counseling	X			
Readiness	X			
Relocation assistance	X			
Unit-sponsored volunteer group				

Service Branches Have Assignment Policies to Mitigate PCS-Related Disruptions

Each service branch has its own set of assignment policies that enhance family stability. Although these assignment policies exist to mitigate PCS move family disruptions, data from the service branches suggest that few service members exercise these policies. Here, we summarize the common programs shared across the service branches. Detailed information about each service branch's assignment policies can be found in Appendix D. These descriptions are based on information provided by each service branch to the project sponsor and are not necessarily exhaustive of all possible assignment policies available to alleviate PCS move disruptions.

Table 3.3
Existing Programs and Services That Address Second-Order PCS Disruptions

Program	Service Member and Spouse Military Life	Service Member and Spouse Psychosocial Outcomes	Child Psychosocial Outcomes	Child School Performance and Engagement	Family Functioning
HEAT					
MilitaryChildCare.com					
Military Family Life Counseling		X	X	X	X
MilitaryINSTALLATIONS					
Military Kids Connect			X	X	
Military OneSource	X	X	X	X	X
TO and Move.mil					
Plan My Move					
SECO					
Military and Family Readiness Centers					
Child and youth programs			X	X	
Employment assistance					
EFMP			X		
Financial counseling					
Readiness	X	X			X
Relocation assistance					
Unit-sponsored volunteer group	X	X			

1. *Basic Allowance for Housing (BAH) waivers:* The Air Force and Marine Corps report that a service member can apply for a BAH waiver to make the BAH allowance based on the dependents' location instead of the service member's permanent duty station in the event that the family lives separately.
2. *Designated location:* Certain service members serving overseas can move their dependents to a designated location of their choosing.
3. *Dual-military spouses:* The service branches attempt to co-locate dual military spouses so they can live in the same household when possible.

4. *EFMP:* The availability of required EFMP services (e.g., medical, educational, early intervention) at duty locations is taken into consideration when assigning members with an EFMP family member.
5. *Extension of overseas tours:* There are various programs that allow service members to extend overseas tours.
6. *High school senior stabilization:* This program allows flexibility in PCS moves to allow for a child to graduate high school.
7. *Home-basing:* This program reduces PCS costs and promotes family stability by allowing certain service members to be assigned back to the previous CONUS location after serving their current overseas tour.
8. *Humanitarian assignments:* Eligible service members experiencing a serious family hardship may be reassigned to a different location or postpone a PCS move.
9. *Time-on-station waivers:* Each service member is required to fulfill minimum time-on-station requirements. Waivers, which allow service members to reduce the time required to serve at a duty station, are granted on a case-by-case basis.

Both the House and the Senate have proposed bills to provide additional flexibility for military families during PCS moves.[3] Both bills would allow certain military families to move during the relocation period, which is defined as the 180 days before or after a service member relocates to his or her new duty station. Eligible military families would include those who meet at least one of the following conditions at the beginning of the relocation period: spouses who are employed or enrolled in a degree, certificate, or license-granting program; children enrolled in school; EFMP families; and family members caring for an immediate family member with a chronic or long-term illness. These proposals would give service members and military families the option to stay or move into temporary government or government-subsidized housing and provide an opportunity for the BAH to be calculated based on the dependent location instead of the service member location.

Summary

DoD and the service branches currently provide a broad spectrum of services to assist service members and their families undergoing a PCS move. Our crosswalks depicted in Tables 3.2 and 3.3 indicate that existing DoD policies, programs, and services address PCS disruptions. However, we do not know the effectiveness of these current services and whether there are certain subgroups of families who would benefit from

[3] Descriptions in this report are based on the May 17, 2017, version of Senate Bill S. 1154 (U.S. Congress, 2017b) and the January 4, 2017, version of House Bill H.R. 279 (U.S. Congress, 2017a). On December 12, 2017, the President signed H.R. 2810, National Defense Authorization Act for Fiscal Year 2018 (U.S. Congress, 2017c).

these services but do not participate. In the next chapter, we discuss scope for future work to investigate the effectiveness of existing services to better understand whether changes could be made to better serve military families and alleviate disruptions associated with PCS moves.

CHAPTER FOUR

Policy Implications and Recommendations

In this report, we provided a comprehensive overview of the types of family disruptions created by PCS moves and conducted a more in-depth analysis of certain disruptions of particular interest to the project sponsor, including spouse employment, retention intentions, and the disruptions related to whether military families move as one unit or move separately at different times. Our analysis demonstrates that moving is disruptive for many families, and our interviews suggest that the service branches understand that certain groups may be disproportionately adversely impacted, including those with family members in the EFMP and employed spouses. We also find that there are some benefits to moving, based on evidence from the literature and from our interviews with SMEs. While there is extensive evidence that PCS moves are associated with spouse employment and retention intentions, there is no readily available information about the extent to which service members and their families move as a unit versus sequentially and the corresponding impact on PCS-related disruptions.[1] Further, much of the evidence linking PCS moves with adverse spouse employment outcomes and retention intentions cannot be interpreted as causal evidence.

Policy Implications

DoD and the service branches currently provide a broad suite of policies, programs, and services aimed at mitigating family disruptions associated with a PCS move. Based on the disruptions identified in Chapter Two and the list of existing programs discussed in Chapter Three, we do not find definitive evidence that new policies, programs, or services are needed to address PCS disruptions.

[1] A reviewer noted that the travel voucher forms that members use to document moving expenses might theoretically be used to gain insight about the frequency of family unit versus sequential moves, as these forms should contain the dates corresponding to when each family member incurred move-related costs. Such analysis would likely be costly and prone to error, and none of our interviewees mentioned this potential data source. Nevertheless, future research might explore whether travel voucher data could be used to assess the frequency of sequential moves.

However, there are opportunities to improve the PCS move process to reduce disruptions, particularly on the front end. Specifically, given survey and interview data related to the timeliness of receiving PCS orders, **we suggest that there is potential for increasing the lead time given to families prior to a PCS move and identify three policy implications to improve the PCS process**. In particular, we recommend that DoD implement the following changes:

- **Identify ways to ensure that funding needed to support PCS moves is available when needed.** When a service member is notified of a PCS move determines how much time that service member and his or her family have to plan their move. Therefore, increasing the lead time of notification before a move is scheduled to occur could help alleviate some of the disruptions and might further assist EFMP families, who need to deal with additional setup costs, such as establishing medical care and educational assistance at the new duty station. Based on data obtained directly from the Navy and information provided through the interviews with SMEs, notification of a PCS move is currently directly tied to funding, and congressional funding and decreasing resources for PCS support programs were the most frequently mentioned issues related to funding PCS moves. This means that, for example, the earlier funding is released by Congress, the sooner service members can be notified of a move and start planning their relocations. While this is not the only source of delay, it is one potential source of timing issues. Thus, DoD should consider any and all programming and budgeting actions that would functionally increase lead time to service members facing a PCS move.
- **Improve the demand signal for logistical aspects of PCS moves to mitigate disruptions.** Our interview analysis suggests that improving the demand signal (defined as the indicator that a move is imminent for those involved in the logistics chain) is particularly important for moves occurring during peak season. The Navy, for example, uses a letter of intent as a trigger for the PCS move process instead of the receipt of funded orders. Earlier identification of a potential move (perhaps before funding is available) would allow families to get a head start in securing movers, establishing transportation, finding housing, and planning other aspects of the move, which could help to alleviate disruptions related to time constraints. While this may benefit the service member, earlier planning could result in additional disruptions for the service member (and the stream of moves used to backfill that position) if the final orders do not match those used for the letter of intent.
- **Sync personnel assignment, financial, and PCS systems electronically.** Currently, the service branch systems that process service member pay and assign ments and those systems that initiate the logistics required for a PCS move (e.g., the Defense Personal Property System [DPS]) do not "talk" to each other. This lack of coordination may cause delays in planning for moves. We suggest insti-

tuting an automatic notification system that would send an electronic alert to TRANSCOM to start the move process. This notification system would likely need to work through a financial system (i.e., personnel actions would impact pay actions and trigger financial transfers, signaling the start of the logistical PCS process). The Marine Corps Total Force System (MCTFS) and the Integrated Personnel and Pay System–Army (IPPS-A) are current and future mechanisms, respectively, where such notification could be tested. The MCTFS is an integrated system, but it does not currently offer automatic notifications.[2] IPPS-A is intended to be an online human resources hub that provides integrated personnel, pay, and talent management capabilities across the total Army. It is expected to be self-service and near real-time, reduce process time, and allow for one-time data entry.[3]

There are a few caveats to keep in mind when considering changes to PCS move policies. The first is that a change to solve one potential disruption could impact another disruption. For example, to minimize school disruptions, PCS moves could be primarily scheduled during summer breaks. However, as we discussed earlier, the peak season for moves is the summer, and scheduling more PCS moves during that time could potentially increase the difficulty that has already been documented surrounding finding available movers. The second caveat is that some disruptions would require changes to long-standing DoD policy. For instance, current DoD assignment policy prevents DoD from taking spouse employment status into account during a service member's assignment process, with few exceptions. Such policy could, however, be changed, after a thorough impact assessment was completed.

Recommendations for Future Study

Further work is needed to understand the effectiveness of current services provided by DoD and the service branches. Additional research should evaluate the relative effectiveness and efficiency of different types of service provision mechanisms (e.g., online, call centers, in person) and should investigate why certain families participate in existing programs and others do not. We also suggest that DoD and the services collect data on the extent to which service members and families move as a unit to determine whether either approach alleviates or exacerbates PCS disruptions and to evaluate proposals designed to give families more flexibility along this dimension.

---

[2] In our interview with representatives from the Office of the Under Secretary of Defense for Acquisition, Technology, and Logistics (OUSD[AT&L]), we were told, "There's a change request for us to receive automated information on PCS moves within the Defense Personal Property System from the Navy and Marine Corps."

[3] For more information on IPPS-A, see www.ipps-a.army.mil/.

Better Understand the Effectiveness of Different Types of Access Points for PCS-Related Support Services

Little is known about the relative effectiveness of different types of service provision mechanisms. In particular, according to our interviewees, funding for in-person support services continues to decline, and many military families report that they prefer to receive relocation and program information from online sources (e.g., Facebook, Twitter). The 2015 ADSS asked active-duty military spouses about their preferred method to learn about military support programs and services and the degree of difficulty in accessing information related to military life. Published tabulations show that 28 percent of military spouses preferred to learn about services through their spouse, 28 percent preferred to learn through websites, and 21 percent preferred to learn through social media (e.g., Facebook, Twitter, LinkedIn). Less preferred methods to learn about programs and services included military family support groups (6 percent), the unit commander (5 percent), on-base family assistance centers (5 percent), and newspapers (2 percent).[4] Figure 4.1 shows that 14 percent of military spouses reported that accessing information related to military life was difficult or very difficult in 2015, while 54 percent reported that it was easy or very easy.

Figure 4.1
How Easy or Difficult Is It for You to Access Information Related to Military Life?

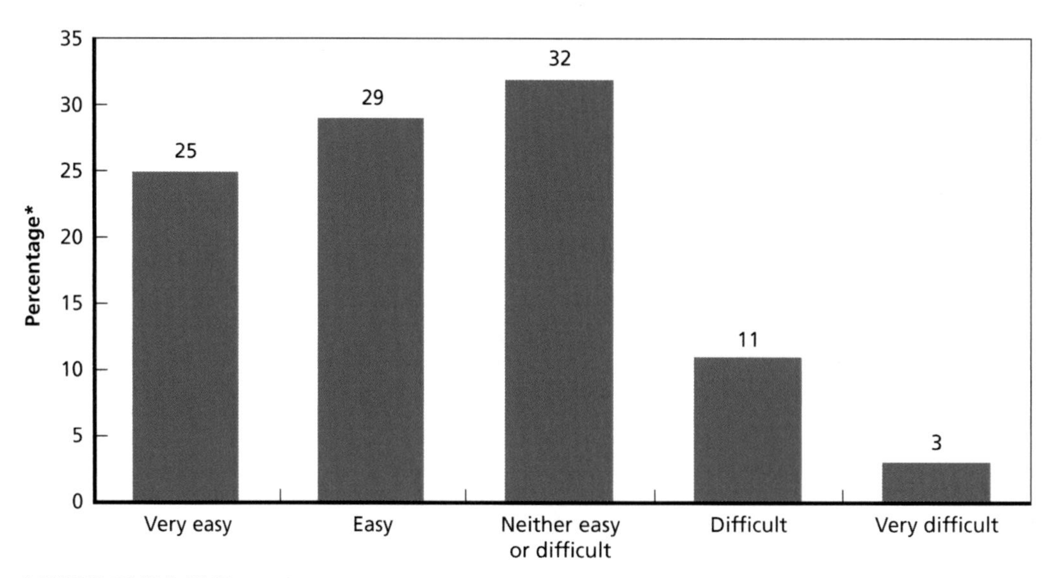

SOURCE: DMDC, 2015.
* Percentage of respondents who answered the relevant survey item.
RAND *RR2304-4.1*

4 These statistics are restricted to those who answered the relevant survey item.

Further, the second–most commonly mentioned way to alleviate PCS disruptions, based on interviews with SMEs, was to use call centers, help lines, social media, and websites. This potentially reinforces a preference to obtain information through access points that do not require an in-person visit to a local office. We also acknowledge that preferences for certain access points may change over time, depending on the quality of information provided at these access points (e.g., funding cuts to in-person services could reduce the quality of information provided). However, this does not necessarily mean that all in-person mechanisms should be replaced by web-based resources or call centers. For example, it could be the case that information about programs may be disseminated more broadly and more cost-effectively through online resources but that in-person access is still needed to troubleshoot certain types of problems or would be more beneficial to certain types of families versus others (e.g., families experiencing their first PCS). Understanding the relationships between access points, including the degree to which they are substitutes versus complements, could have implications for ways to better serve military families. Future work should determine whether certain programs are redundant and identify ways to improve program efficiency.

Better Understand When and Why Families Do and Do Not Use PCS-Related Programs, Services, and Assignment Policies

Service members and family members actively choose whether to participate in or utilize PCS-related programs and services and PCS-related assignment policies. Yet it is not clear what underlies the motivation behind these decisions. For example, data from the service branches suggest that very few families request a waiver to delay a PCS move, but we do not know whether low usage is due to a lack of awareness, negative stigma (e.g., fear of how it will appear to leadership), potential loss of a desired assignment, or some other factor. Further, certain families choose to participate in certain programs because they know they will directly benefit from participating, while those who do not participate might not be aware of such programs or may actively decide not to participate because they have other means of support outside of the programs offered. Future work is needed to determine whether there are certain subgroups of participants who would like to participate in an existing program but currently do not participate and how to reduce barriers that these families may face to participating.

Identify Ways to Collect Data Related to Whether Families Move as a Unit or Separately

DoD generally, and the service branches specifically, currently cannot identify how many military families move as a unit versus separately (e.g., service member moves first, family moves first). As a result, we do not know how frequently each type of move occurs and whether either approach mitigates or exacerbates PCS disruptions. Given that Congress has put forth proposals to give military families more options to move separately,[5] it is important to collect such data to determine the extent to which families already decide to move sequentially, to estimate the potential size of the population that could be affected by such proposals, and to evaluate its impact on PCS disruptions.

Summary

While we did not find any obvious gaps in existing resources available to military families undergoing PCS moves, we have suggested ways to improve the front end of the move process by (1) increasing the amount of notice service members and families receive prior to a relocation to provide more time to plan for the move process itself, (2) determining a better way to trigger logistical aspects of the move in a timely fashion, and (3) synchronizing assignment and PCS systems to improve coordination.

Further, we do not yet know the effectiveness of the existing resources in alleviating PCS-related disruptions. Therefore, we recommend that future research be pursued to better understand the effectiveness of different access points to these services (e.g., online, call centers, in person) and to determine why families choose to participate in programs. We also recommend that DoD and the service branches collect data on the extent to which military families move as a unit versus separately so that an analysis can be done on whether either method mitigates or exacerbates PCS disruptions and so that the potential impact of proposed legislation that would allow for additional flexibility along this dimension could be estimated.

---

[5] S. 1154 and H.R. 279 were introduced by the Senate and House, respectively, in 2017 as part of the 115th Congress.

APPENDIX A

Interview Methodology

As part of our mixed-methods approach, we conducted semistructured interviews with program personnel and SMEs. Below, we discuss the process by which we selected interview participants, interviewed them on various aspects of the PCS process, and drew preliminary conclusions from the information they provided. Figure A.1 depicts the components of our interview methodology. The study methodology was reviewed by RAND's Human Subjects Protection Committee and was determined not to be human subjects research.

Limitations on the Number and Type of Interviews Conducted

The quick-turn nature and condensed timeline for this study precluded our team from seeking the necessary approvals for conducting a more extensive number of interviews. Expanding our list to more than nine interviews with civilians (i.e., not uniformed military personnel) would have required approval from the Office of Management and Budget (OMB), as stipulated in the Paperwork Reduction Act from 1995. While this is less problematic in protracted studies that span the majority of the fiscal year (FY), in streamlined studies, the time alone needed for OMB approval can inadvertently derail the data collection trajectory. Thus, the decision to conduct fewer than ten interviews obviated the need for OMB approval and complemented our condensed study schedule.

Selection of Personnel and SME Interviewees

Being limited to nine interviews necessitated that we closely coordinate with the project sponsor to select an ideal cross-section of participants who, despite being few in number, could afford us a fulsome appreciation of the challenges related to PCS moves. The condensed nature of the study also precluded us from interviewing service members from the active and reserve components who had personally undergone a PCS move, which would have required additional time for Human Subjects Protection Committee approvals. We settled on a small but substantive group of interview participants (described further below), some of whom were DoD personnel and management experts and others of whom were service branch–specific SMEs on the PCS process and the issue of retention.

Figure A.1
Interview Methodology

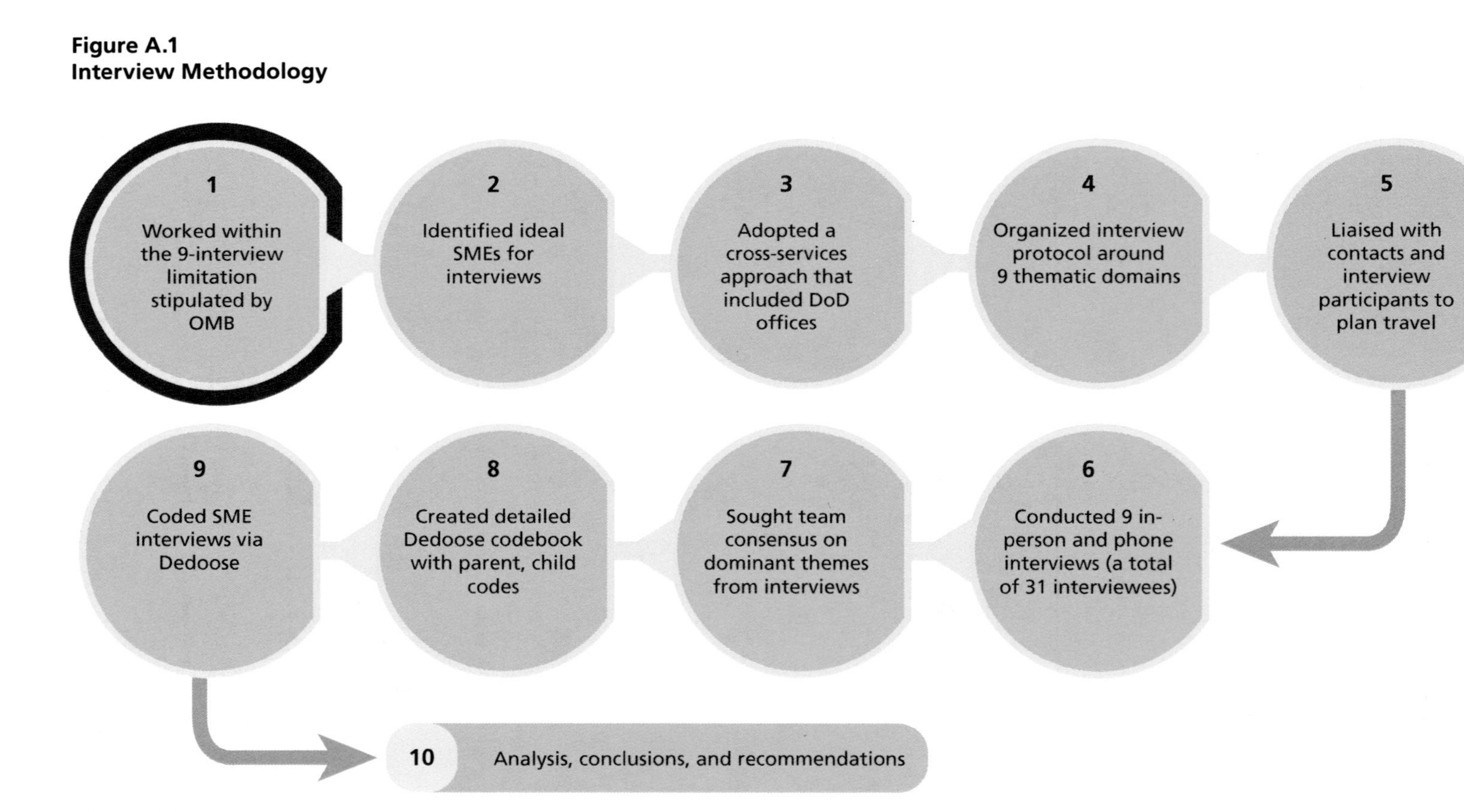

RAND *RR2304-A.1*

Employing a Cross-Service Branch Approach

Recognizing that each service branch might approach the PCS process differently, we understood the value in a cross-sectional approach that utilized SMEs from the Air Force, Army, Marine Corps, Navy, and Coast Guard. We also spoke with representatives from several DoD offices, including DTMO, OUSD(P&R), and OUSD(AT&L).[1]

We found that each of these DoD offices had tailored expertise, in addition to general information about the PCS process. For example, the purview of the DTMO is to manage allowances provided to service members through the Joint Travel Regulations, such as with housing allowances, temporary lodging expenses, and the shipping of personally operated vehicles. We likewise learned that OUSD(P&R) provides policy oversight for the family readiness system, including the family support centers in both the active and reserve components, and oversees the DoD spouse education and development opportunities program. OUSD(AT&L) had a narrow focus that centered on the movement of personal property, including developing policy for the movement of household goods. Table A.1 shows the full range of interviews conducted, including the participating office or organization and the number of interviewees per organization.

Focusing on Nine Thematic Domains

As mentioned above, two of our data collection methods included reviewing past literature and reviewing secondary data sources on the PCS process. After doing so, the information gaps that remained compelled us to develop a semistructured interview

Table A.1
Organizations That Participated in Interviews

Service or Department	Organization	Number of Participants	Date
U.S. Marine Corps	Manpower and Reserve Affairs	9	June 13, 2017
USCG	Headquarters	6	June 14, 2017
DoD	DTMO	1	June 14, 2017
Department of the Army	Headquarters (Pentagon)	2	June 15, 2017
Department of the Navy	Deputy Chief Naval Operations N1	7	June 15, 2017
DoD	OUSD(P&R)	2	June 15, 2017
DoD	OUSD(AT&L)	3	June 16, 2017
U.S. Air Force	Headquarters (Pentagon)	1	June 19, 2017
Total number of interviewees: 31			

[1] This study was conducted prior to the reorganization proposed in the FY 2017 National Defense Authorization Act (Public Law 114–328, 2016).

protocol with questions that would add fidelity to what we already knew and provide the missing information we sought. The semistructured approach allowed us to employ open-ended questions that elicited broad and expansive responses from our SMEs, further providing the topical nuance we needed. We then sorted the interview questions into nine categorical domains: (1) general disruptions associated with PCS moves; (2) negative aspects of PCS moves; (3) positive aspects of PCS moves; (4) timeliness of PCS orders; (5) retention challenges associated with the PCS process; (6) general aspects of the PCS assignment process; (7) assessing the problem and impact of PCS moves; (8) existing policies, programs, and services; and (9) money and funding related to the PCS process.[2] Most questions were followed by secondary questions, which were sub-questions or probing questions that helped facilitators guide the interview's trajectory and make ample use of the allotted time. Box A.1 depicts our SME interview protocol.

Coordinating Interview Schedule with Participant

As described above, we worked with the project sponsor to identify ideal SMEs with whom we should speak and reached out to a range of individuals across the different service branches and within DoD. Due to the truncated project timeline, we strove to conduct the in-person interviews over a three-day period in mid-June 2017, with an outlying interview occurring a few days later. With the exception of our interview with a DTMO representative, which we conducted via telephone, and our interview with an OUSD(AT&L) representative, which we conducted at our Virginia-based RAND office, all interviews were held at the respective buildings of the participating organizations or offices.

Conducting In-Person Interviews with SMEs

Because some organizations and offices included more participants than originally planned, our nine interviews netted a total of 31 interviewees. For example, our interview with the Marine Corps Office of Manpower and Reserve Affairs included nine participants, our interview with the office of the Deputy Chief of Naval Operations N1 included seven participants, and our interview with the headquarters element of the USCG included six participants. Each conversation lasted an average of one hour and began by (1) introducing our research team and the purpose of the study, (2) reviewing the informed consent statement, and (3) asking participants to comment on their background vis-à-vis the PCS process. We also underscored the voluntary nature of the interviews. At the onset of each conversation, we reinforced to participants that their

[2] The nine thematic domains are not mutually exclusive, and the same protocol question often elicited responses that crossed domains. For example, interview questions and probes nested under the money and funding domain often applied to the timeliness of PCS orders domain (e.g., congressional funding directly impacts the timely issuance of PCS orders) and the negative aspects of PCS moves domain (e.g., personal costs are incurred during moves).

Box A.1
SME Interview Protocol

1. **Background information**
   a. What is your current position?

   b. How long have you been in your current position?

   c. Can you briefly describe your roles and responsibilities?

2. **Disruptions associated with PCS moves**
   a. What specific PCS-related issues has your office or service branch identified as most disruptive (negative) for military family stability, health, and well-being?
      i. Or, what potentially negative consequences of PCS moves has your office or service branch focused on?
      ii. How do these impacts vary by family member (e.g., service member, spouse, school-aged child, other family member)?
      iii. Can you tell us more about how PCS moves may impact a spouse's employment and career opportunities?
         1. Can you tell us more about how PCS moves may impact retention (across all family members)?
         2. How frequently do these disruptions materialize?

   b. What, if any, *positive* effects of PCS moves has your office or service branch identified?

   c. What specific issues, both positive and negative, are associated with not receiving timely PCS orders (i.e., experiencing a short period of time between notification and the actual move)?
      i. Do you have any data on average lead time and changes in lead time over time (e.g., across FYs)? Can you share that with us? [Share Navy example.]

   d. What specific issues, both positive and negative, are associated with receiving PCS orders during specific times of the year (e.g., summer, fall, holiday season)?

   e. What data sources do you use to determine consequences—either positive or negative—associated with PCS moves?
      i. Do you have any data on tiered migration? That is, when a service member [and] his or her family do not move together, at the same time, but one relocates to the new installation before (or after) the other. Where does this data come from? Would you be willing to share it with us?

identities would be safeguarded. Given this, when taking meeting notes, we referred to participants numerically rather than addressing them by name. For example, in our transcript from the Office of Manpower and Reserve Affairs interview, the nine participants were referred to as R1, R2, R3 . . . R9. We concluded each conversation by explaining the study timeline and asking whether we could reach out to participants with follow-on questions, if needed.

Building Consensus for Dominant Themes

Our interview analysis began shortly after data collection concluded. The process was trifurcated into three steps: building team consensus on the most salient points raised in interviews, constructing a highly detailed codebook for use during the qualitative coding exercise, and coding the interview responses. For the first of the three steps, a subset of team members selected two interview transcripts, reviewed them in full, derived a list of eight frequently mentioned themes, and solicited feedback on their saliency from the rest of the team. We conceived of these themes as macro-level categorical "bins" or "parent codes," under which we could nest micro-level "child codes" that allowed for more-nuanced categorization.

Creating a Dedoose Codebook with Parent and Child Codes

Following the identification of salient themes, our second step included the development of a codebook. We created an Excel spreadsheet that contained the parent and child codes described above. Central to this typology were inclusion and exclusion rules that clarified how to categorize the full gamut of interview responses for team members assisting with the coding effort. Figure A.2 shows a select portion of this typology, including the inclusion and exclusion rules.

Coding Interview Responses

For the third step in our data analysis process, we used Dedoose qualitative coding software to import interview transcripts, organize this source material, and code participant responses. The Excel-based codebook helped to guide this process. To assess coding reliability at the onset, two different team members coded the same set of interview transcripts to ensure that they were "binning" respondents' statements in a fashion commensurate with the inclusion and exclusion rules in the codebook. Following this exercise, coding decisions were discussed, amendments to the codebook were made, and the remainder of interview transcripts were coded. Ultimately, 813 statements or excerpts from our nine interviews were binned under eight parent codes and 63 different child codes, as depicted in Figure A.3.[3] While it is too extensive to depict here, the Dedoose software also showed the number of interview excerpts that were binned per each of the 63 child codes.

Derived Conclusions and Policy Implications

We concluded the coding effort by charting the aggregated interview data for each of the eight themes. These quantitative bar charts provided a user-friendly visual of the frequency of interview excerpts per each child code. We added orange callout boxes to highlight the excerpts with high frequency rates. The charts assisted our project team in developing conclusions, policy implications, and recommendations for future study. Figures A.4 and A.5 provide two such examples of frequency charts and their associated conclusions.

[3] Note that we had nine categorical domains for our interview questions but developed eight parent codes during our Dedoose coding exercise. This was because we folded the "General Disruptions" category into the "Negative Aspects of PCS Move" parent code.

Figure A.2
Screen Shot of a Portion of PCS Moves Codebook

04 Retention (Parent Code)	Child Code	Inclusion Rule	Exclusion Rule
	04.01 Confirming and denying there is a retention problem	Issues related to: any conversation or excerpt where someone claims the presence or absence of a retention problem--focus on general statements; any explanation of the retention problem	Exclude: information on explicit actions that have been taken to rectify the retention problem and reduce attrition (such as the granting of waivers and provision of allowances)--instead code as 04.03 (Waivers, exceptions, allowances and other fixes)
	04.02 Frequency of PCS moves as impactful	Issues related to: PCS moves, especially the frequency of moves, impacting retention; any mention of retention bonuses and their ability to help, or not; any mention of the PCS process affecting recruiting efforts in the future; any mention of military recruiting being low compared to years past	Exclude: anything not related to how PCSing might impact retention and/or recruitment
	04.03 Waivers, exceptions, allowances and other fixes	Only issues related to: waivers, exceptions and allowances within the context of retention; also include other explicit actions that have been taken to rectify the retention problem and reduce attrition	No explanation needed
	04.04 Reserves and National Guard	Issues related to: Reserves and National Guard (for all Services) and how the PCS process might differ for them	Exclude: all information pertaining to active duty forces and special operation forces
	04.05 Other	No explanation needed	No explanation needed
05 Reassignment Process			
	05.01 Delay requests from servicemember	Issues related to: the process by which servicemembers request to move at a later date; problems associated with such requests; ability of the service in question to grant such requests	Exclude: any information related to the process by which the service or service-specific office informs the servicemember that he or she will be PCSing; when the PCS notification process is either adequate or
	05.02 Value of geographic stabilization and reassignment as last resort (keeping people in place)	Issues related to: how the specific service in question, or DoD writ large, tries hard not to move servicemembers and their family; how allowing a family to remain in place and not move is often beneficial	No explanation needed

RAND *RR2304-A.2*

Figure A.3
Coding Results for Interviewee Excerpts

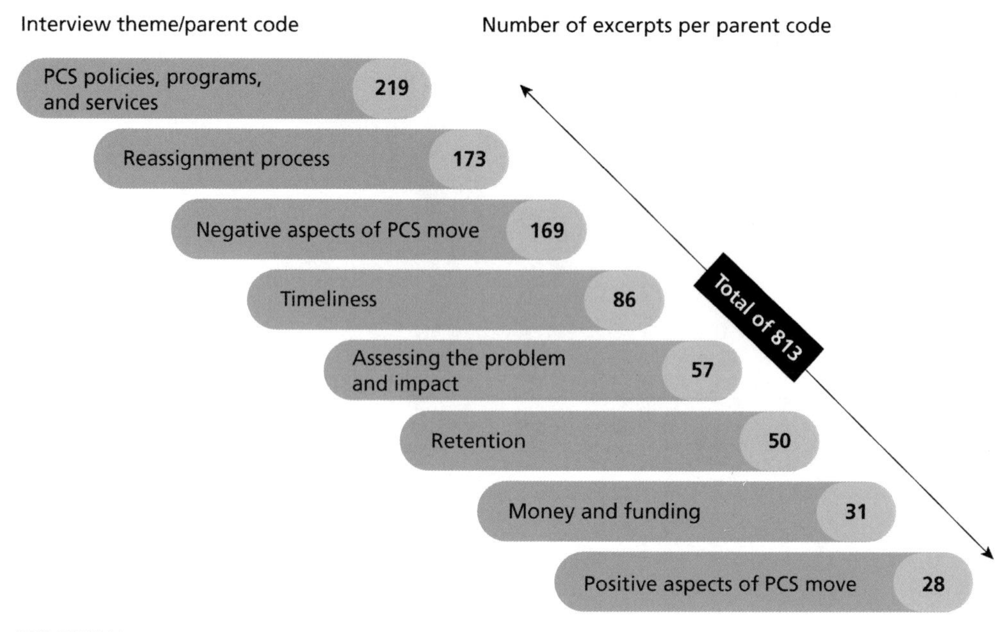

RAND *RR2304-A.3*

Figure A.4
Charted Interview Data and Conclusions: Positive Aspects of PCS Move

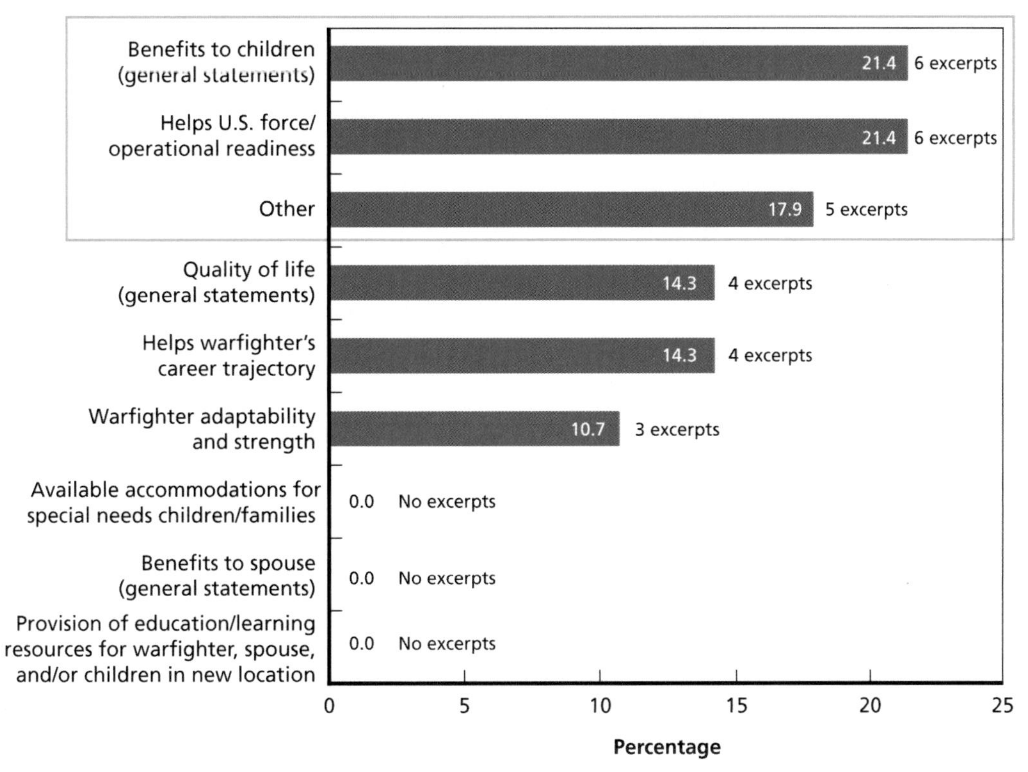

RAND *RR2304-A.4*

Figure A.5
Charted Interview Data and Conclusions: Reassignment Process

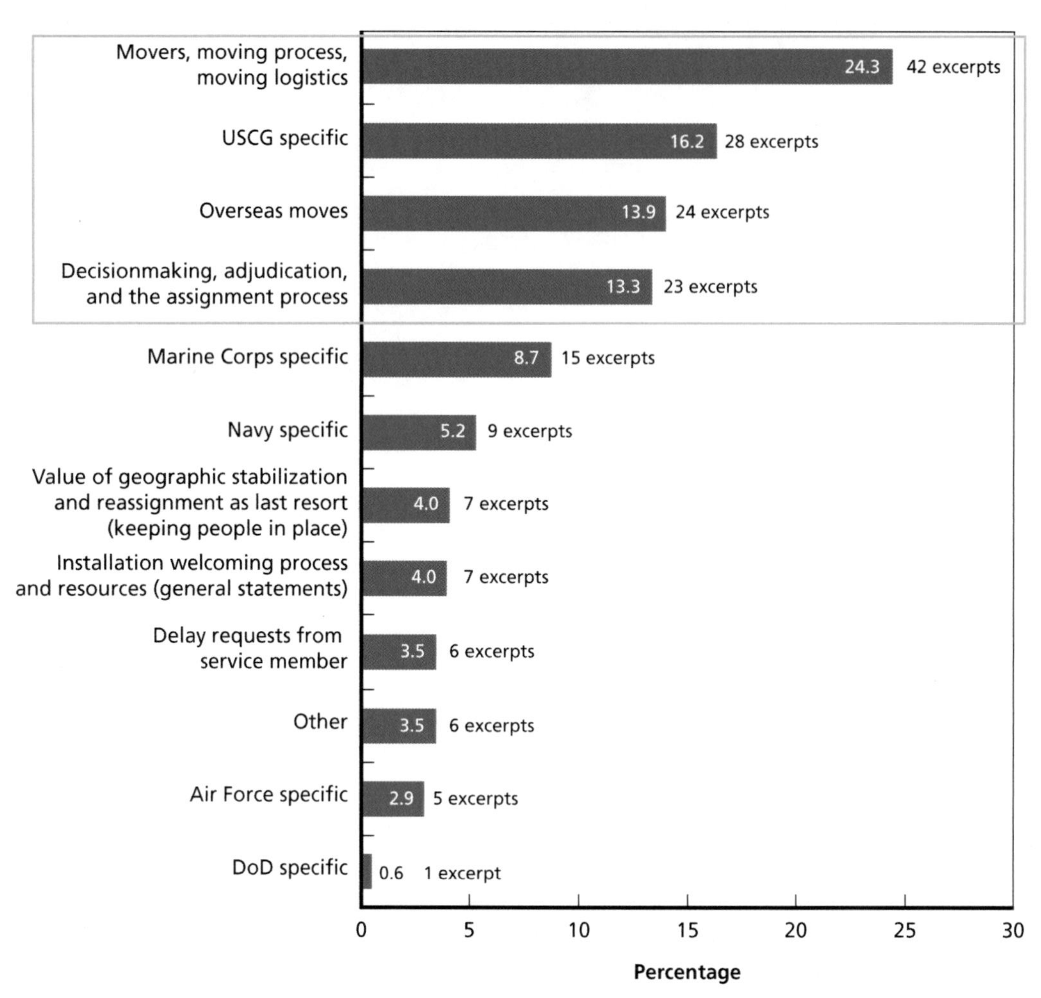

RAND *RR2304-A.5*

APPENDIX B

Deployment Life Study Analysis

In this appendix, we describe in detail the DLS analysis conducted to study how service member, spouse, and teen attitudes about military life and job satisfaction change during the PCS move cycle.

About the Deployment Life Study

The DLS is a longitudinal study of roughly 2,700 married military families across all branches and both components. The study was designed to follow married families through a complete deployment cycle (pre-, during, and post-) to identify antecedents, correlates, and consequences of family readiness (see Tanielian et al., 2014). The survey also includes information about the timing and frequency of PCS moves. The data were collected from 2011 through 2015, and each family member was surveyed every four months for three consecutive years. The DLS was designed to be representative of all *married* military families (with and without children) and thus is not representative of all military families. Additional restrictions during analysis (e.g., including only married families that experienced a PCS move) further limits generalizability.

The analysis examines four outcomes potentially related to retention behavior:

- *Financial stress* (service member and spouse). This measure assesses respondents' current financial condition, difficulty paying bills, ability to save money, and concern about their current financial situation. Higher values indicate higher levels of stress.
- *Commitment to the military* (service member, spouse, and teen). This measure includes three items assessing the degree to which being part of the military inspires the respondent to do the best they can as a service member, spouse, or military child; the degree to which the respondent is willing to make sacrifices to be in the military; and the degree to which the respondent is glad they, their spouse, or their parent is part of the military. Higher values indicate greater commitment.

- *Satisfaction with military life* (service member and spouse). These measures vary slightly by respondent, but all assess how much the respondent is satisfied with their quality of life as a service member or military spouse. Higher values indicate greater satisfaction.
- *Retention intentions* (service member, spouse, and teen). These measures also vary by respondent, but all assess how likely the service member is to remain on active duty (should he or she have the choice) or how much the spouse or teen favors the service member remaining in the military. Higher values indicate greater likelihood or favoritism of remaining in the military.

Analytic Methods

We used growth curve models with linear splines to describe outcomes over the course of the time period around a PCS move for multiple respondent types in the family (i.e., service member, spouse, and child), including short-term and long-term periods before and after a move. PCS information is derived from survey questions that capture the month and type of all pre-baseline moves and any new moves during the study. This information is used to calculate a continuous month of the move. The month of the move is time varying, with 0 indicating the exact month of the move, positive values indicating months after the move, and negative values indicating months preceding the move. The analyses were limited to the subsample of families who experienced at least one PCS move during the period of observation. For the handful of individuals having more than one PCS move during the study, we censored observations after the first move.

These analyses are descriptive and are not interpreted as causal effects of PCS moves on outcomes. Further, the DLS data do not contain information about when service members were first notified about their relocation, meaning that some service members might have known a move was imminent, while others might not have. The months until the start of a PCS move were recoded into six linear splines and three step functions (more details below). This technique was used to allow both slopes and intercepts to vary. Linear splines for months until the start of a study deployment included knots at –6 months, –2 months, 0 months, +2 months, and +6 months, thus defining six periods with linear trajectories in each: (1) more than six months prior to the PCS move, (2) between six months and two months prior to the move, (3) between two months prior to the PCS move and the move, (4) between the move and two months post-move, (5) between two months and six months post-move, and (6) more than six months after the PCS move. The time periods captured by the six splines are summarized in Figure B.1. We also include steps at –2 months, 0 months, and +2 months from the PCS move.[1]

Wald tests were used to assess two statistical hypotheses to test whether the trajectory around the PCS move was different from what we would expect to see by chance:

1. Does the overall trajectory differ significantly from a flat line (i.e., no changes in average values over the move cycle)? This was tested using a joint Wald test in which all trajectory factors are tested against a null hypothesis that the parameters = 0.
2. Does the overall trend differ significantly from a straight line (i.e., a constant change over time that is unrelated to the phase of, or timing within, the PCS move period)? This was tested using a slightly different base model that included a linear change over time unrelated to PCS moves. We then performed a joint Wald test against a null hypothesis that all trajectory factors = 0.

Table B.1 presents the p values for these two tests. The outcomes with asterisks designate those in which the trend around the PCS move is different from *both* a flat or straight line. In particular, we found that service member commitment to the military, satisfaction with military life, retention intentions (marginally significant), and spouse financial stress around a PCS move were significantly different from both a flat and a straight line.

Because these tests did not tell us anything about the direction of the trends, we plotted the predicted outcomes over time. Because several correlated parameters were used to model outcome trajectories, the model parameters themselves are not easily interpretable as descriptions of the trajectories. To convert these parameters into a useful description, we used the model to derive a predicted average trajectory within our sample for a specific, hypothetical person's PCS move based on service member (for service member outcomes) and spouse (for spouse outcomes) reports of the month

Figure B.1
PCS Period Timeline for DLS Trajectory Analysis

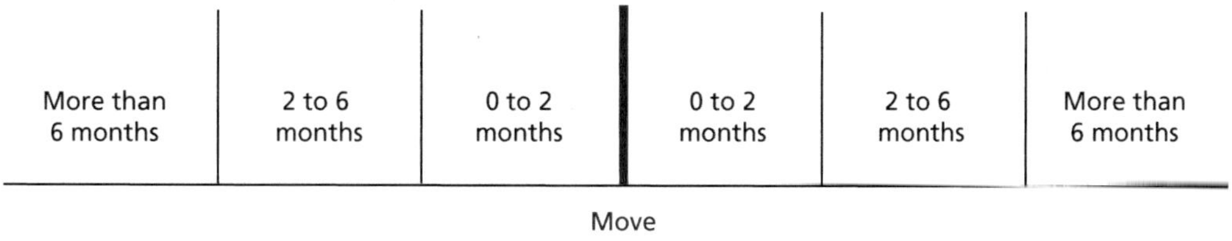

More than 6 months	2 to 6 months	0 to 2 months	0 to 2 months	2 to 6 months	More than 6 months

Move

RAND *RR2304-B.1*

[1] For each model, we also ran joint Wald tests to test whether the steps were jointly equivalent to 0. If so, then we used a simplified version of the model: one with slopes only. Most of the models did not require the additional steps (changes in level), with two exceptions: service member military satisfaction and spouse financial stress.

Table B.1
***P* Values from Global Statistical Tests for Deployment Life Study Analysis**

Type of Respondent	Outcome	*P* Values for Test 1	*P* Values for Test 2
Service member	Financial stress scale	0.3005	0.4735
	Commitment to military	0.0039*	0.0168*
	Satisfaction with military life	0.0291*	0.0440*
	Retention intentions	0.0274*	0.0762*
Spouse	Financial stress scale	0.0000*	0.0016*
	Commitment to military	0.2466	0.4591
	Satisfaction with military life	0.5050	0.7021
	Retention intentions	0.5865	0.3034
Teen	Commitment to military	0.5641	0.4058
	Retention intentions	0.4275	0.1949

* *P* values are statistically significant at the 10-percent level or higher.

NOTE: Test 1 examines whether the overall trajectory differs significantly from a flat line; Test 2 examines whether the overall trajectory differs significantly from a straight line.

of their PCS move. The particular hypothetical PCS cycle selected was designed to be relatively typical and to capture short-term and long-term patterns of outcomes around PCS moves. The cycle covers a one-year period beginning six months before a PCS move and following individuals for six months after the move occurs. The two-month period before and after the move reveals more-immediate changes, whereas the six-month period gives us some indication of baseline functioning before a move and longer-term functioning post-move.[2] Also, note that we are not necessarily capturing *couples* in our analysis; that is, even though the DLS data are composed only of married couples, both members of the couple did not always participate in each survey wave. These plots are depicted in Figures 2.6 through 2.9 and discussed in Chapter Two.

---

[2] To some extent, the selection of the periods used in the trajectory model is arbitrary, but it should be guided by theory or existing data. Our decision to use six months as the longer-term period was based on information from the Navy. In data provided to the sponsor, the Navy indicated that the Chief of Naval Operations has established a goal of an average of six months of lead time for sailor notification of an upcoming PCS.

APPENDIX C

Detailed Descriptions of Programs Addressing PCS Disruptions

This appendix contains detailed descriptions of programs provided by DoD (and included in Table 3.1) and service branch–specific programs addressing PCS disruptions.

Detailed Descriptions of Programs Provided by DoD

Housing Early Assistance Tool

HEAT is an online system that can be used to obtain housing assistance prior to a PCS move. Assistance includes access and information on private housing, military base housing managed by the installation's Housing Services Office, rental partnership programs, and community housing. Rental partnership programs are community properties that have a formal agreement with the installation's Housing Services Office and offer discounted and/or flexible housing options to military families.

MilitaryChildCare.com

MilitaryChildCare.com is a website that provides comprehensive information on child care programs that are operated through the military or military-approved. Military families create an online profile and have access to a search engine that allows them to search for child care options.

Military Family Life Counseling Program

The Military Family Life Counseling Program offers nonmedical counseling and briefings on and off base to service members and their families. These counselors provide guidance on a broad range of issues, such as stress management, family relationships, and crisis intervention. The program also offers presentations that cover subjects common to most military families, like deployments and reintegration. Child and youth behavioral military and family life counselors are also available as part of this program and are meant to help children address a wide array of issues, including those that occur during PCS moves (e.g., changes at home, family and school relationships).

MilitaryINSTALLATIONS

MilitaryINSTALLATIONS is a searchable website that provides installation-specific information about the location, cost of living, and population, as well as resources available at each installation, such as relocation assistance, child care, housing, and family programs. MilitaryINSTALLATIONS also includes installation-specific PCS resources. For example, some installations have lending lockers, which are programs to lend household items, such as dishware, televisions, and futons, to military families going through a PCS move. In addition, certain installations also offer some amount of free child care to military families immediately before or after a PCS move.

Military Kids Connect

Military Kids Connect is an online community for military children aged 6 through 17 that provides resources on ways to cope with stressors related to military life. The website also includes information about higher education for teenagers, as well as information for parents, caregivers, and teachers.

Military OneSource

Military OneSource is a centralized website that provides information on service member and military family–related resources. Categories covered on Military OneSource include the military life cycle, family and relationships, moving and housing, financial and legal, education and employment, health and wellness, recreation, travel and shopping, and service providers and leaders. In terms of relocation assistance specifically, the website provides general tips, factsheets, and podcasts about relocation, as well as links to other resources, such as Plan My Move and Move.mil. Military OneSource also provides contact information for emergency financial assistance for each service branch, which could be used if out-of-pocket expenses from a PCS move are greater than anticipated. The website also has links to community-specific information, such as cost of living, crime, community comparison, school report cards, salary analyzers, job networks, and a relocation finance calculator. To assist with issues related to children moving, Military OneSource provides information covering issues related to changing schools, child development programs, and children with special needs. Service members and their spouses can create a personal profile on Military OneSource to obtain information most relevant to their personal situation. Military OneSource offers its own confidential nonmedical counseling and also provides information about and access to the Military and Family Life Counseling Program.

Transportation Office and Move.mil

Each installation has a TO that offers counseling with information about the move process and government entitlements. Service members are required to obtain counseling before making a government move. This counseling can be conducted in person at a local TO, or, for certain types of moves, service members can use Move.mil to self-counsel and facilitate the logistics of the move through the website. Eligibility to use Move.mil varies by service branch. For example, some service branches allow first-time movers to self-counsel using Move.mil, while others do not.

The TO also assists with the logistics of the move, including the scheduling and shipment of household goods, property, pets, and vehicles, and a housing office that can help families determine their housing allowances, arrange temporary housing, and provide information on housing options.

Plan My Move

Plan My Move is an online tool that creates a calendar with PCS move–related tasks listed by day. Service members and their families input information about their move dates and locations, and their personalized calendar is created.

Spouse Education and Career Opportunities Program

The SECO program provides military spouses with a broad range of resources to facilitate the pursuit of education and careers. There are several components of the SECO program.

1. The MySECO website provides information about SECO resources and includes interactive features to help military spouses plan their careers. In terms of education resources, MySECO contains search engines for colleges and scholarships. To assist with military spouse employment, MySECO includes a resume builder, an occupation research tool, and an individual career plan builder.
2. The Military OneSource Career Center offers certified career counseling to cover career exploration, education, training and licensing information, employment readiness, and career opportunities. SECO also provides bundled counseling services that cover specific fields, such as health care and intelligence and security, and specific topics, such as reentering the workforce.
3. The Military Spouse Employment Partnership (MSEP) is a program that identifies and creates partnerships with companies who agree to promote military spouse careers. Partners sign a Statement of Support with the Armed Forces that commits them to identify and promote employment opportunities of military spouses, post job listings and their human resources pages on the MSEP Career Portal, offer portable jobs to relocating military spouse employees, and mentor new MSEP partners. Military spouses can search for job opportunities and identify partners through the MSEP Career Portal.
4. The My Career Advancement Account (MyCAA) Scholarship is a program administered by DoD that provides financial assistance of up to $4,000 to support the pursuit of portable careers among eligible military spouses. To be eligible, a military spouse must be married to an active-duty service member in pay grades E-1 to E-5, W-1 to W-2, or O-1 to O-2 and must be able to begin

and finish their coursework while the service member is on Title 10 military orders. The military spouse cannot be in the Armed Forces. The financial assistance could be used to pay for education and training courses required to obtain an associate degree in an approved field. Approved fields cover a broad range of occupations in areas including business, education, hospitality, health care, information technology, and a variety of skilled trades.[1] In addition, the assistance might be used to pay for an occupational license or certification, including location-based licenses or credentials in approved fields.

Detailed Descriptions of Service Branch–Specific Programs

Air Force

Airman and Family Readiness Centers

The Airman and Family Readiness Centers (A&FRCs) provide a broad range of services to promote military family readiness, including relocation assistance, counseling, and education. Each A&FRC gives relocation assistance through consultations, workshops, and outreach activities. Prior to a family arriving at the local installation, these centers provide the family members with a relocation package to ease the transition. Upon arrival to the local installation, families receive a newcomer brief that provides local information. Many centers have a lending closet that provides household goods to families on a temporary basis. The A&FRCs offer families with EFMP status additional support during PCS moves, including access to on- and off-installation resources and state and federal resources. EFMP coordinators from these centers work with families to address concerns regarding assimilation.

The A&FRCs Heart Link program is an orientation program for Air Force spouses who are relatively new to active duty (five years or less). The program provides services and resources to help new military spouses integrate into the military community in order to strengthen military families, enhance readiness, and promote retention.

Army

Army Community Service Centers

The Army Community Service (ACS) centers are available at every installation and provide a broad range of services to service members and their families, including relocation assistance. For example, ACS centers provide counseling to assist families undergoing a PCS move and information about the destination installation. Orientations are provided to families when they first arrive at an installation to inform them of services, such as lending closets, and to help them acclimate to the new destination.

---

[1] See MyCAA Scholarship (undated) for a comprehensive list.

Orientations are also provided to service members who are moving to an overseas location, and these orientations are open to family members.

Army OneSource

Army OneSource is similar to Military OneSource. It provides information about many of the same resources listed on Military OneSource. Additionally, it contains information about Army-specific programs and resources.

Marine Corps

Marine Corps Community Services

Marine Corps Community Services (MCCS) consist of a broad range of programs covering relocation assistance, family readiness, children and youth, child care, and military spouse employment assistance. The MCCS website (www.usmc-mccs.org) is a centralized website that provides general information about the programs offered by MCCS, as well as detailed information about specific services at each installation.

MCCS offers distant counseling, which makes counseling services accessible to military families who are not located near a base (e.g., when displaced from a PCS move).

MCCS has an interactive acculturation program to promote adjustment and family function called Lifestyle Insights, Networking, Knowledge, and Skills (L.I.N.K.S.). L.I.N.K.S. offers workshops tailored to each type of family member (e.g., spouses, teens, Marines, school-aged children, Marine parents). Topics covered include familiarization with Marine Corps history, local installation services, military pay and benefits, issues related to relocation, and separation and deployment.

MarineOnLine

MarineOnLine provides family care training modules to build family resiliency during PCS moves. Modules include ways for children to maintain friendships when they move and how to be involved with the moving process.

DSTRESS Line

DSTRESS Line is a confidential counseling service for service members and their families that is available 24 hours a day, seven days a week. This service is provided through phone, chat, and Skype.

Navy

Fleet and Family Support Centers

Navy Fleet and Family Support Centers (FFSCs) are installation-specific centers that provide a range of services to service members and their families. As of October 1, 2015, DoD stopped funding relocation programs at the service branch level. To reduce costs, FFSCs provide service members and their families referrals to services by giving them access to a network of providers, such as the TO, public affairs offices, Military

OneSource, Navy Supply Systems Command Household Goods, Navy Personal Property, the Exceptional Family Member Program, Navy Housing, Navy and Child Youth Programs, and school liaison officers. To further facilitate transitions to new installations, FFSCs have a sponsor program. Certain installations also offer lending lockers to newly arriving military families. Similar to the other service branches, the Navy Child and Youth Programs offer child care, school transition assistance, and youth and teen recreational activities. Half of the employees of these Child and Youth programs are military spouses. Military spouse employees can be transferred to their next duty station with the same grade and pay without needing to reapply for a position.

Coast Guard

CG SUPRT Program

The CG SUPRT Program provides information about military and family readiness programs and offers 24/7 support via phone. The program provides confidential counseling that covers a broad range of topics, including marital and family problems, financial consultation, spouse career education, and moving and relocation. The program offers up to 12 sessions per issue per year and is available to Coast Guard active-duty members, civilian employees, members of the Selected Reserve, and their family members.

APPENDIX D

Service Branch–Specific Assignment Policies Related to PCS Moves

Below, we provide additional information on assignment policies related to PCS moves that was provided by each service branch. Differences in programs across service branches may reflect actual differences in the assignment policies offered or differences in the amount of detail provided by each service branch.

Air Force

The Air Force has the following PCS move assignment policies to mitigate disruptions to certain families:

1. BAH Waiver: The service member's BAH is based on the permanent duty location. The service member may request a secretarial BAH waiver to receive a BAH based on the dependents' location instead of the new permanent duty location to allow children to complete a school year, to account for the member being on extended temporary duty en route to the new location, and to allow a member to complete schooling (less than a year) that requires a PCS move.
2. Consecutive Overseas Tour (COT) and In-Place Consecutive Overseas Tour (IPCOT): A COT occurs when a service member completes an outside the continental United States (OCONUS) tour and carries out a follow-on OCONUS assignment right after the first one is completed. An IPCOT occurs when a service member serves a consecutive tour at the same OCONUS location.
3. Designated Location: Service members serving unaccompanied overseas tours can move their dependents to any location within CONUS.
4. Dual Spouse Military: Air Force members with a spouse on active duty are eligible for Join Spouse assignment consideration, which would allow the couple to live in the same household.
5. EFMP: The EFMP assignment policy places the service member in a location based on personnel requirements and the availability of required EFMP services (e.g., medical, educational, early intervention).

6. High School Senior Assignment Deferral: This program allows a PCS move to be deferred to allow a dependent child to complete the senior year of high school. The use of this policy is determined case by case and applies only to active-duty officers at the O-5 level or below and enlisted members at the E-8 level or below.
7. Home-Basing/Follow-on Assignments: This program allows certain service members to be assigned back to the same CONUS location or allows consideration of an advanced assignment. The purpose of this program is to reduce PCS move costs and to increase family stability.
8. Humanitarian Assignment: The Air Force allows reassignment and deferment of service members who need to locate in or near a specific location to address short-term hardships involving a family member. A humanitarian assignment typically occurs once to address the short-term problem. Circumstances that warrant a humanitarian assignment include a death of a spouse or child within 12 months (includes miscarriages occurring after 20 weeks or more in gestation), serious financial problems,[1] abandonment of children while serving an unaccompanied tour in which the children cannot relocate to the overseas location, a family member who is terminally ill and not expected to live beyond two years, and when an authorized state or local agency puts a child into the service member's home and a deferment is necessary to comply with local laws to complete an adoption.
9. Time-on-Station (TOS) Waivers: For CONUS-to-CONUS moves, the TOS standard is 48 months. For CONUS-to-OCONUS moves, the TOS is 12 months for first-term service members and 24 months for career service members. TOS waivers are given on a case-by-case basis.

Army

The Army has the following PCS move assignment policies to mitigate disruptions to certain families:

1. Compassionate Assignments Program: This program allows service members to be reassigned or temporarily deferred from an assignment to address a family hardship, including a serious illness or death in the family.
2. COT and IPCOT: A COT occurs when a service member completes an OCONUS tour and carries out a follow-on OCONUS assignment right after

[1] Serious financial problems cannot be a result of personal overextension of financial resources. To qualify for a humanitarian assignment, the financial problem cannot be resolved through the service member going on leave, correspondence, power of attorney, or any other method that does not require a relocation or deferment.

the first one is completed. An IPCOT occurs when a service member serves a consecutive tour at the same OCONUS location.

3. EFMP: The EFMP assignment policy places the service member in a location based on the availability of required EFMP services (e.g., medical, educational, early intervention) and stabilizes families by giving the service member four-year assignments.
4. High School Senior Stabilization Program: This program allows PCS moves to be scheduled in such a way as to mitigate school year disruptions for members with children in their junior and senior years of high school.
5. Home-Basing/Advanced Assignment Program: This program applies to enlisted members who complete 12-month dependent-restricted tours. Under this program, these members could be assigned back to the same CONUS location after completing their tours or may be given an advanced assignment to a new duty location. For those with advanced assignments, the Army tries to move members to one of their preferred locations when possible. The purpose of this program is to reduce PCS move costs and to increase family stability.
6. Married Army Couples Program: Army members with a spouse on active duty are eligible for assignment consideration, which would allow the couple to live in the same household (with the distance between the two assignments defined as a 50-mile radius or one hour of driving time).
7. TOS Waivers: There is a 36-month TOS requirement for CONUS assignments. Waivers and exemptions must be approved by the Secretary of the military department or by an approved authority delegated by the Secretary.

Marine Corps

The Marine Corps has the following PCS move assignment policies to mitigate disruptions to certain families:

1. BAH Waiver: A service member on a CONUS assignment may request a waiver so that the BAH is based on the dependents' location in the event that the service member and the family decide to live in separate households.
2. COT and IPCOT: A COT occurs when a service member completes an OCONUS tour and carries out a follow-on OCONUS assignment right after the first one is completed. An IPCOT occurs when a service member serves a consecutive tour at the same OCONUS location.
3. Designated Place: Members with OCONUS PCS orders can opt to have their dependents locate to a different location: a *designated place*. The member must choose the designated place prior to the execution of the PCS orders. This includes members with orders to dependent-restricted OCONUS locations and

accompanied OCONUS locations at which the member chooses to serve unaccompanied.

4. Dual-Military Spouse: The Marine Corps will try to assign dual-military spouses to the same location, but there is no explicit policy to do so like the other service branches.
5. EFMP: The Marine Corps reviews PCS move orders of EFMP families to determine whether required EFMP services (e.g., medical, educational, early intervention) are available at the new duty location.
6. High School Continuity (Continuity of Education [COE]): Dependents might be authorized to delay their move until an extended break in the school year or at the end of the current academic year under certain circumstances dictated by the Marines Corps regulations. For example, members with a high school senior dependent can make this request so that the dependent can complete the school year. Only one request for COE for a high school senior dependent is allowed.
7. Home-Basing: This program assigns Marines back to the same CONUS location after completing dependent-restricted OCONUS tours.
8. Humanitarian Transfers: This program allows service members with short-term (defined as 36 months or less) family hardships to transfer to a different station or to cancel a PCS move to remain at the current station.
9. Overseas Tour Extension Incentive Program (OTEIP): The goal of this program is to promote the stability of personnel overseas. Those who extend their tours under OTEIP are not eligible for COT or IPCOT travel allowances, and vice versa.
10. Single Parents: A service member who becomes a single parent due to unforeseen circumstances, such as the death of a spouse, can apply to defer a tour or to be reassigned. A single parent may defer an assignment to a dependent-restricted area or to an accompanied tour where the dependent travel is denied for six months after the date of delivery (single mothers only) or after the adoption of the child.
11. TOS Waivers: There is a 36-month TOS requirement for CONUS assignments. TOS waivers are approved on an individual basis.

Navy

The Navy has the following PCS move assignment policies to mitigate disruptions to certain families:

1. COT and IPCOT: A COT occurs when a service member completes an OCONUS tour and carries out a follow-on OCONUS assignment right after

the first one is completed. An IPCOT occurs when a service member serves a consecutive tour at the same OCONUS location.

2. Designated Place: Members with OCONUS PCS orders may opt to have their dependents locate to a different location: a *designated place*. The member must choose the designated place prior to the execution of the PCS orders. This includes members with orders to dependent-restricted OCONUS locations and accompanied OCONUS locations at which the member chooses to serve unaccompanied.
3. Dual Military Couples/Joint Household: Dual military couples can opt into the program once so that the Navy will co-locate couples when possible for all future moves. This is a recent change to Navy assignment policies.
4. EFMP: The EFMP assignment policy places the service member in a location based on the availability of required EFMP services (e.g., medical, educational, early intervention).
5. Home-Basing: Home-basing includes no-cost PCS moves, which occur when a member's new assignment is within the existing permanent duty location. The Navy stated that it is attempting to increase the number of no-cost PCS moves not only to reduce the costs associated with moving but also to increase the amount of time a dependent resides in any given location.
6. Humanitarian Assignments: Members might be eligible for a humanitarian assignment due to family hardships, as dictated by Navy regulations.
7. Overseas Tour Extension Incentives Program: Eligible enlisted members may extend their overseas tour voluntarily by 12 months or more and receive an incentive (e.g., rest and relaxation time, lump-sum payment).
8. Stabilization for Sailors with High School Seniors: Service members with a child in the junior year of high school can request to stay in the same location until the child graduates.
9. TOS Waivers: There is a 36-month TOS requirement for CONUS assignments. TOS waivers are given for various reasons, including to fulfill readiness requirements and support career advancement.

References

Angrist, Joshua D., and John H. Johnson, "Effects of Work-Related Absences on Families: Evidence from the Gulf War," *Industrial and Labor Relations Review*, Vol. 54, No. 1, 2000, pp. 41–58.

Bourg, Chris, and Mady Wechsler Segal, "The Impact of Family Supportive Policies and Practices on Organizational Commitment to the Army," *Armed Forces and Society*, Vol. 25, No. 4, 1999, pp. 633–652.

Burke, Jeremy, and Amalia R. Miller, "The Effects of Job Relocation on Spousal Careers: Evidence from Military Change of Station Moves," *Economic Inquiry*, 2017.

Burrell, Lolita M., Gary A. Adams, Doris Briley Durand, and Carl Andrew Castro, "The Impact of Military Lifestyle on Demands on Well-Being, Army, and Family Outcomes," *Armed Forces and Society*, Vol. 33, No. 1, 2006, pp. 43–58.

Castaneda, Laura Werber, and Margaret C. Harrell, "Military Spouse Employment: A Grounded Theory Approach to Experiences and Perceptions," *Armed Forces and Society*, Vol. 34, No. 3, 2008, pp. 389–412.

Chandra, Anita, Sandraluz Lara-Cinisomo, Lisa H. Jaycox, Terri Tanielian, Rachel M. Burns, Teague Ruder, and Bing Han, "Children on the Homefront: The Experience of Children from Military Families," *Pediatrics*, Vol. 125, No. 1, 2010, pp. 16–25.

Chandra, Anita, Laurie T. Martin, Stacy Ann Hawkins, and Amy Richardson, "The Impact of Parental Deployment on Child Social and Emotional Functioning: Perspectives of School Staff," *Journal of Adolescent Health*, Vol. 46, No. 3, 2010, pp. 218–223.

Chartrand, Molinda M., Deborah A. Frank, Laura F. White, and Timothy R. Shope, "Effect of Parents' Wartime Deployment on the Behavior of Young Children in Military Families," *Journal of Pediatric Adolescent Medicine*, Vol. 162, No. 11, 2008, pp. 1009–1014.

Clever, Molly, and David R. Segal, "The Demographics of Military Children and Families," *The Future of Children*, Vol. 23, No. 2, 2013, pp. 13–39.

Defense Manpower Data Center, *November 2003 Status of Forces Survey of Active Duty Members: Tabulations of Responses*, Ft. Belvoir, Va.: Defense Research, Surveys, and Statistics Center, DMDC Report No. 2003-03, 2004.

Defense Manpower Data Center, *December 2005 Status of Forces Survey of Active Duty Members: Overview Briefing*, Ft. Belvoir, Va., 2006.

Defense Manpower Data Center, *December 2007 Status of Forces Survey of Active Duty Members Topics: Permanent Change of Station (PCS) Moves and Details on Readiness*, presentation, Ft. Belvoir, Va., 2008. As of April 30, 2018:
http://www.esd.whs.mil/Portals/54/Documents/FOID/Reading%20Room/Personnel_Related/2007-12%20SOFS-A_Briefing_PCS.pdf

Defense Manpower Data Center, *2013 Status of Forces Survey of Active Duty Members: Tabulations of Responses*, Ft. Belvoir, Va.: Defense Research, Surveys, and Statistics Center, DMDC Report No. 2013-050, 2014.

Defense Manpower Data Center, *2015 Survey of Active Duty Spouses: Tabulations of Responses*, Ft. Belvoir, Va.: Defense Research, Surveys, and Statistics Center, DMDC Report No. 2015-028, 2015.

Defense Manpower Data Center, *February 2016 Status of Forces Survey of Active Duty Members: Tabulations of Responses*, Ft. Belvoir, Va.: Defense Research, Surveys, and Statistics Center, DMDC Report No. 2016-035, 2017.

Defense Manpower Data Center Military Family Life Project, *Active Duty Spouse Study, Longitudinal Analyses 2010–2012, Project Report*, Office of the Deputy Assistant Secretary of Defense for Military Community and Family Policy, 2015.

DMDC—*See* Defense Manpower Data Center.

DoD Instruction 1342.22, *Military Family Readiness*, July 3, 2012, incorporating Change 2, April 11, 2017. As of August 19, 2018:
http://www.esd.whs.mil/Portals/54/Documents/DD/issuances/dodi/134222p.pdf

Drummet, Amy Reinkober, Maryilyn Coleman, and Susan Cable, "Military Families Under Stress: Implications for Family Life Education," *Family Relations*, Vol. 52, No. 3, 2003, pp. 279–287.

Engel, Rozlyn C., Luke B. Gallagher, and David S. Lyle, "Military Deployments and Children's Academic Achievement: Evidence from Department of Defense Education Activity Schools," *Economics of Education Review*, Vol. 29, No. 1, 2010, pp. 75–82.

GAO—*See* U.S. General Accounting Office.

Hanushek, Eric A., John F. Kain, and Steven G. Rivkin, "Disruption Versus Tiebout Improvement: The Costs and Benefits of Switching Schools," *Journal of Public Economics*, Vol. 88, 2004, pp. 1721–1746.

Harrell, Margaret C., Nelson Lim, Laura Werber, and Daniela Golinelli, *Working Around the Military: Challenges to Military Spouse Employment and Education*, Santa Monica, Calif.: RAND Corporation, MG-196-OSD, 2004. As of April 30, 2018:
https://www.rand.org/pubs/monographs/MG196.html

Heaton, Paul, and Heather Krull, *Unemployment Among Post-9/11 Veterans and Military Spouses After the Economic Downturn*, Santa Monica, Calif.: RAND Corporation, OP-376-OSD, 2012. As of April 30, 2018:
https://www.rand.org/pubs/occasional_papers/OP376.html

Hengstebeck, Natalie D., Sarah Meadows, Beth Ann Griffin, Esther M. Friedman, and Robin Beckman, "Military Integration," Chapter Seven in Sarah O. Meadows, Terri Tanielian, and Benjamin R. Karney, eds., *The Deployment Life Study: Longitudinal Analysis of Military Families Across the Deployment Cycle*, Santa Monica, Calif.: RAND Corporation, RR-1388-A/OSD, 2016, pp. 263–302. As of April 30, 2018:
https://www.rand.org/pubs/research_reports/RR1388.html

Hosek, James, Beth Asch, Christine Fair, Craig Martin, and Michael Mattock, *Married to the Military: The Employment and Earnings of Military Wives Compared with Those of Civilian Wives*, Santa Monica, Calif.: RAND Corporation, MR-1565-OSD, 2002. As of April 30, 2018:
https://www.rand.org/pubs/monograph_reports/MR1565.html

Hosek, James, Beth J. Asch, and Michael G. Mattock, *Should the Increase in Military Pay Be Slowed?* Santa Monica, Calif.: RAND Corporation, TR-1185-OSD, 2012. As of April 30, 2018:
https://www.rand.org/pubs/technical_reports/TR1185.html

Hosek, James, and Shelley MacDermid Wadsworth, "Economic Conditions of Military Families," *The Future of Children*, Vol. 23, No. 2, 2013, pp. 41–59.

Krauss, B. M., *The Impact of Changes in Geographic Mobility on the Wages of the Military Family Between 1985 and 1992*, Wright-Patterson Air Force Base, Ohio: Air Force Institute of Technology, No. AFIT/GCA/LAR/96S-8, 1996.

Lakhani, Hyder, "Reenlistment Intentions of Citizen Soldiers in the U.S. Army," *Armed Forces and Society*, Vol. 22, No. 1, 1995, pp. 117–130.

Lakhani, Hyder, and Stephen. S. Fugita, "Reserve/Guard Retention: Moonlighting or Patriotism?" *Military Psychology*, Vol. 5, No. 2, 1993, pp. 113–125.

Lavee, Yoav, Hamilton I. McCubbin, and Joan M. Patterson, "The Double ABCX Model of Family Stress and Adaptation: An Empirical Test by Analysis of Structural Equations with Latent Variables," *Journal of Marriage and Family*, Vol. 47, No. 4, 1985, pp. 811–825.

Lester, Patricia, Hilary Aralis, Maegan Sinclair, Cara Kiff, Kyung-Hee Lee, Sarah Mustillo, and Shelley MacDermid Wadsworth, "The Impact of Deployment on Parental, Family, and Child Adjustment in Military Families," *Child Psychiatry and Human Development*, Vol. 47, No. 6, 2016, pp. 938–949.

Lim, Nelson, Daniela Golinelli, and Michelle Cho, *"Working Around the Military" Revisited: Spouse Employment in the 2000 Census Data*, Santa Monica, Calif.: RAND Corporation, MG-566-OSD, 2007. As of April 30, 2018:
https://www.rand.org/pubs/monographs/MG566.html

Lim, Nelson, and David Schulker, *Measuring Underemployment Among Military Spouses*, Santa Monica, Calif.: RAND Corporation, MG-918-OSD, 2010. As of April 30, 2018:
https://www.rand.org/pubs/monographs/MG918.html

Lyle, David S., "Using Military Deployments and Job Assignments to Estimate the Effect of Parental Absences and Household Relocations on Children's Academic Achievement," *Journal of Labor Economics*, Vol. 24, No. 2, 2006, pp. 319–350.

Maury, Rosalinda M., and Brice Stone, *Military Spouse Employment Report*, Syracuse, N.Y.: Institute for Veterans and Military Families, 2014. As of May 5, 2017:
https://ivmf.syracuse.edu/article/military-spouse-employment-survey/

McCubbin, Hamilton I., and Joan M. Patterson, "The Family Stress Process: The Double ABCX Model of Adjustment and Adaptation," *Marriage and Family Review*, Vol. 6, No. 1–2, 1983, pp. 7–37.

Meadows, Sarah O., Beth Ann Griffin, Benjamin R. Karney, and Julia Pollak, "Employment Gaps Between Military Spouses and Matched Civilians," *Armed Forces and Society*, Vol. 42, No. 3, 2016, pp. 542–561.

Meadows, Sarah O., Terri Tanielian, and Benjamin R. Karney, eds., *The Deployment Life Study: Longitudinal Analysis of Military Families Across the Deployment Cycle*, Santa Monica, Calif.: RAND Corporation, RR-1388-A/OSD, 2016. As of April 30, 2018:
https://www.rand.org/pubs/research_reports/RR1388.html

Mohr, Debrah A., Robert L. Holzbach, and Robert F. Morrison, *Surface Warfare Junior Officer Retention: Spouses' Influence on Career Decisions*, San Diego, Calif., USN-Naval Personnel Research and Development Command Technical Report NPRDC 81-17, 1981.

Morgan, Mary E., *Permanent Change of Station and Stress*, U.S. Army War College Military Studies Program Paper, 1991.

MyCAA Scholarship, "Career Search," undated. As of September 12, 2018:
https://mycaa.militaryonesource.mil/mycaa/Career/Search.aspx

Orthner, Dennis K., *Family Impacts on the Retention of Military Personnel*, Alexandria, Va.: U.S. Army Research Institute for Behavioral and Social Sciences, Research Report 1556, 1990.

Public Law 114–328, National Defense Authorization Act for Fiscal Year 2017, December 23, 2016. As of September 20, 2018:
https://www.gpo.gov/fdsys/pkg/PLAW-114publ328/content-detail.html

Richardson, Amy, Anita Chandra, Laurie T. Martin, Claude Messan Setodji, Bryan W. Hallmark, Nancy F. Campbell, Stacy Hawkins, and Patrick Grady, *Effects of Soldiers' Deployment on Children's Academic Performance and Behavioral Health*, Santa Monica, Calif.: RAND Corporation, MG-1095-A, 2011. As of April 30, 2018:
https://www.rand.org/pubs/monographs/MG1095.html

Rockoff, Jonah E., and Benjamin B. Lockwood, "Stuck in the Middle: Impacts of Grade Configuration in Public Schools," *Journal of Public Economics*, Vol. 94, No. 11–12, 2010, pp. 1051–1061.

Rosen, Leora N., and Doris Briley Durand, "The Family Factor and Retention Among Married Soldiers Deployed in Operation Desert Storm," *Military Psychology*, Vol. 7, No. 4, 1995, pp. 221–234.

Schwartz, Amy Ellen, Leanna Stiefel, and Sarah A. Cordes, "Moving Matters: The Causal Effect of Moving Schools on Student Performance," *Education Finance and Policy*, Vol. 12, No. 4, 2017, pp. 419–446.

Schwartz, Amy Ellen, Leanna Stiefel, Ross Rubenstein, and Jeffrey Zabel, "The Path Not Taken: How Does School Organization Affect Eighth-Grade Achievement?" *Educational Evaluation and Policy Analysis*, Vol. 33, No. 3, 2011, pp. 293–317.

Schwerdt, Guido, and Martin R. West, "The Impact of Alternative Grade Configurations on Student Outcomes Through Middle and High School," *Journal of Public Economics*, Vol. 97, 2013, pp. 308–326.

Shiffer, Cristin O., Rosalinda V. Maury, Hisako Sonethavilay, Jennifer L. Hurwitz, H. Christine Lee, Rachel K. Linsner, and Michella S. Mehta, *2017 Blue Star Families Military Family Lifestyle Survey: Comprehensive Report*, Encinitas, Calif.: Blue Star Families and the Institute for Veterans and Military Families, 2017. As of November 15, 2017:
https://bluestarfam.org/survey/

Spencer, Emily, Kimberley Page, and Matthew G. Clark, "Managing Frequent Relocation in Families? Considering Prospect Theory, Emotional Framing, and Priming," *Family and Consumer Science Research Journal*, Vol. 45, No. 1, 2016, pp. 77–90.

SteelFisher, Gillian K., Alan M. Zaslavsky, and Robert J. Blendon, "Health-Related Impact of Deployment Extensions on Spouses of Active Duty Army Personnel," *Military Medicine*, Vol. 173, No. 3, 2008, pp. 221–229.

Tanielian, Terri, Benjamin Karney, Anita Chandra, and Sarah O. Meadows, *The Deployment Life Study: Methodological Overview and Baseline Sample Description*, Santa Monica, Calif.: RAND Corporation, RR-209-A/OSD, 2014. As of October 7, 2017:
https://www.rand.org/pubs/research_reports/RR209.html

U.S. Chamber of Commerce, *Military Spouses in the Workplace: Understanding the Impacts of Spouse Employment on Military Recruitment, Retention, and Readiness*, Washington, D.C.: Hiring our Heroes Foundation, 2017.

U.S. Congress, 115th Congress, 1st Session, Military Family Stability Act of 2017, Washington, D.C., H.R. 279, January 4, 2017a. As of August 19, 2018:
https://www.congress.gov/bill/115th-congress/house-bill/279/text

U.S. Congress, 115th Congress, 1st Session, Military Family Stability Act, Washington, D.C., S. 1154, May 17, 2017b. As of August 19, 2018:
https://www.congress.gov/bill/115th-congress/senate-bill/1154/text

U.S. Congress, 115th Congress, 1st Session, National Defense Authorization Act for Fiscal Year 2018, Washington, D.C., H.R. 2810, December 12, 2017c. As of August 19, 2018:
https://www.congress.gov/bill/115th-congress/house-bill/2810/text

U.S Department of the Treasury and Department of Defense, *Support Our Military Families: Best Practices for Streamlining Occupational Licensing Across State Lines*, Washington, D.C., 2012.

U.S. General Accounting Office, *Longer Time Between Moves Related to Higher Satisfaction and Retention*, Washington, D.C., GAO-01-841, 2001.